INTRODUCTION TO PLASTICS

Introduction to
PLASTICS

Lionel K. Arnold

THE IOWA STATE UNIVERSITY PRESS, AMES

1968

CHEMISTRY

LIONEL K. ARNOLD, professor of Chemical Engineering at Iowa State University, holds the B.S., M.S., and Ph.D. degrees from Iowa State University. He is a member of the American Institute of Chemical Engineers, the Iowa Academy of Science, Sigma Xi, Phi Kappa Phi, and Phi Lambda Upsilon. Dr. Arnold's chief research has been in dialdehyde starch-gluten polymers and in plastics obtainable from agricultural products and from furfural. Besides this book, he has written numerous technical papers and bulletins for *Modern Plastics, Industrial and Engineering Chemistry, Encyclopedia of Chemical Technology,* and the Iowa State University Engineering Research Institute.

© 1968 The Iowa State University Press
Ames, Iowa, U.S.A. All rights reserved

Composed and printed by
The Iowa State University Press

First Edition, 1968

Second printing, 1969

Standard Book Number: 8148-1272-0

Library of Congress Catalog Card Number: 68-12020

Preface

THE broad and ever-growing field of plastics has produced a great number of books. Many of these are excellent for their intended purpose but hardly seem suitable to the author for use in a one-quarter college course in plastics technology. This book was developed from a belief that such a course should provide a broad general knowledge of the plastics field upon which the student might build in the future to meet his own specific needs. *Introduction to Plastics* was developed from lectures by the author that were refined through his classroom use. It is believed the book will be helpful also to those in business and industry who need to obtain a general view of the plastics industry.

The sources from which the material in this book has been derived are so many and so diverse that it is impractical to attempt to acknowledge them individually. The obvious sources such as books, magazines, and manufacturers' publications and films have been freely consulted. Many people both off and on the campus have been helpful. Men skilled in industrial technology have lectured to his classes and discussed plastics with the author. To all of these he is sincerely grateful. To his wife, for her patience and understanding, go the author's special thanks.

LIONEL K. ARNOLD

Contents

INTRODUCTION TO PLASTICS

Introduction

SOME BASIC CONCEPTS

Plastics mean different things to different people. There are about forty different plastics, each with many uses, which amount to a total production in excess of 13 billion pounds per year in the United States alone (see Table I.1.) They are available in a wide variety of colors as well as black and white. Others are transparent and colorless. They are molded into an endless variety of objects from small knobs, screw caps, and handles to radio and television cabinets. Toys and household items in bright colors are common. Plastic films are used in packaging. Plastic coatings on wood and metal may be both beautiful and resistant to wear and liquids. Other plastics are formed into car and boat bodies. Still others are made into prosaic, chemically resistant pipes and tanks for use in industry. Prices of these plastics vary from as low as 15 cents to as much as $10 a pound (see Table I.2). Somehow we must find a definition to characterize this large group of related materials.

A *plastic*, as defined by the American Society for Testing Materials and the Society of the Plastics Industry, Inc., is "a material that contains as an essential ingredient an organic substance of large molecular weight, is sold in its finished state, and, at some stage in its manufacture or its processing into finished articles, can be shaped by flow." Designating the "es-

Table I.1:   Synthetic Resins and Cellulosics Sales and Production, 1966*

Market	Production	Sales
	million pounds	
Alkyds	642	340
Cellulosics	187	181
Coumarone-indene and		
Petroleum Resins	338	342
Epoxy Resins	145	130
Phenolics	972	783
Polyesters	461	407
Polyethylenes (high density)	920	900
Polyethylenes (low density)	2500	2640
Polypropylenes	510	490
Polystyrene and Styrene		
Copolymers	2310	2400
Polyvinyl Chloride and		
Copolymers	2200	2215
All Other Vinyls	512	415
Ureas and Melamines	654	553
Miscellaneous †	900	830
TOTAL	13,251	12,626

Source: *Modern Plastics Magazine*, January, 1967, p. 115. (Published by McGraw-Hill, Inc., 330 West 42nd Street, New York, N.Y.)
* Data do not include synthetic elastomers and fibers which amount to about 8 billion pounds additional.
† Includes acetal, acrylic, fluoroplastics, nylon, phenoxies, polycarbonate, silicones, and others.

sential ingredient" as "an organic substance" differentiates plastics from ceramic products which are inorganic. Perhaps substitution of "organic high polymer" for "organic substance of large molecular weight" might be more definitive since it would then rule out waxes, which are essentially mixtures of esters of higher saturated monatomic alcohols, and higher fatty acids, both

Table I.2:   Typical Plastics Prices

Plastics	Price Range*
	cents per pound
ABS	33–47
Acetals	65–75
Acrylics	46–56
Polycarbonates	100–170
Cellulosics	40–80
Chlorinated Polyethers	260–350
Polyethylenes	17–23
Epoxies	48–60
Nylons	88–125
Phenolics	20–30
Polypropylenes	25–40
Polystyrenes	15–27
PTFE	550–1000
PVC	16–29

* Approximate. Prices vary with quantity purchased, time, and additives such as color and reinforcements or fillers.

of which are not generally considered as plastics. The shaping by "flow" usually, though not always, involves heat as well as pressure. Certain resins forming essential ingredients in coatings are generally accepted as plastics.

In industry, rubbers, natural and synthetic, usually are traditionally not included in the category of plastics. However, with the similarity between certain synthetic rubbers and certain plastics the distinction seems rather arbitrary. In this discussion rubbers will be included in an effort to complete the overall picture. Modified and synthetic fibers are not always included as plastics. Yet acetate textile fiber is essentially only extruded cellulose acetate plastic. Likewise molded and spun nylon are basically the same material as are cellophane and rayon. These considerations appear to justify at least brief discussion of plastic fibers.

A few general definitions may be helpful. The term *resin* is usually applied to the principal material, generally a high polymer, in the plastic product. This is not always a resinous material, as for example, the cellulose esters and the protein plastic materials. It may be the only constituent but commonly is blended with such additives as fillers, plasticizers, and colorants. A *filler* is usually a solid, such as wood flour, added to the resin to improve the physical properties, such as strength, of the end product. Usually the filler is cheaper than the resin. If the primary purpose of the added material is to cheapen the cost without greatly reducing desirable properties it is usually called an *extender*. *Plasticizers* are solids or liquids, usually organic, added to improve properties, such as flexibility. Fillers are most commonly used with thermosetting materials, plasticizers with thermoplastic materials. *Thermosetting* plastics are those which become permanently hard when heated above a certain temperature. *Thermoplastic* materials are those which soften or liquefy when heated and harden upon cooling. The distinction between thermoplastic and thermosetting materials is becoming less definite. Thermoplastic materials may become thermosetting by addition of small amounts of unsaturated compounds or by being irradiated with certain rays. Thermosetting plastics are being formed by injection and extrusion methods which previously were limited to only thermoplastics. A *molding powder* is a solid mixture of a resin and other materials, such as fillers, suitable for molding. It is frequently not powdered but granular. The term *molding compound* is sometimes used to refer to molding mixtures which are not true compounds.

Perhaps the greatest single advantage of plastics is that they can be molded into the finished products at a relatively low cost compared to the machining and fabricating necessary with wood and metals. Ordinarily the plastic molding materials are more expensive than metal, a higher cost which is frequently offset by the low plastic molding cost. All plastic products are not made entirely by simple molding. Some finishing operations may be necessary. Other plastics may be made into films, sheets, or standard rod or bar stock from which the final products may be fabricated. Large quantities of plastic films are used in packaging and other applications. Plastic resins are becoming of increasing importance in the field of coatings, both decorative and protective.

Other advantages of plastics are inherent in their properties. In general they resist chemicals well and are nonconductors of electricity. High strength-weight ratios are characteristic. They do not rust like iron. On the other hand, they have certain disadvantages which must be considered before applying them to specific uses.

In general, plastics materials include a wide range of products of different properties and costs which are highly competitive between themselves and with many of the older products. Production of these plastics materials has increased rapidly and is expected to continue increasing for many years in the future.

Since *high polymers* form the essential part of all plastics they will be discussed first. This will be followed by a discussion of the general methods used in industry to produce plastic materials such as the molding powders. The methods used for making finished plastic products from the plastics material will then be outlined. Following this, individual plastics will be discussed.

CHAPTER 1

High Polymers

POLYMERIZATION

The general class of chemical reactions by which high polymers are formed is usually referred to as *polymerization*. There are two general types of reactions: *addition* polymerization and *condensation* polymerization. These two can also be combined in the production of a single polymer.

Types of Polymerization

Addition Polymerization. Addition polymerization is a reaction which results in the bonding of two or more molecules without the elimination of any by-product molecules. Two general types of reactions are recognized: (1) those involving a single molecular species and (2) those involving more than one molecular species.

Of the first group the reactions with compounds having carbon to carbon unsaturation are common. Olefinic compounds, as $CR_2{=}CR_2$ where R represents hydrogen or a substituted radical, are the simplest of this type. Examples include ethylene, styrene, vinylidene chloride, and methyl methacrylate. These polymer molecules have the same percentage composition as the monomers from which they are produced. As the double bond opens up, the resulting radicals connect up into long chain compounds. Compounds with triple bonds such as acetylene,

7

HC≡CH, may also react. If only one of the triple bonds opens up, the product is similar to that of the double bond compound except that each segmer contains a double bond. However, two bonds may open up, thus producing a radical which may combine with other radicals at two instead of one bond. Reactions of compounds with carbon to heteroatom unsaturation where another atom not a carbon atom is attached to a carbon atom by at least a double bond, as $R_2C=Z$, also occur. The compounds include aldehydes, ketones, thioaldehydes, and thioketones.

Certain ring compounds may undergo a rupture of the ring followed by polymerization to produce straight chain polymers. These may be homocyclic carbon ring compounds such as cyclopropane, or heterocyclic compounds such as caprolactam.

In the second group are the copolymers. Copolymerization results when two different monomers react to form a polymer at relatively the same energy level and rates. The ratio of the two monomers making the polymer may vary considerably among different copolymers. The arrangement of the monomer units may be random, as XXYXXXXYYXY, or alternating, as XYXYXYXY. *Terpolymers* may be formed from three different monomers. A special type of copolymer known as a block polymer is formed by joining a prepolymerized short chain of one monomer to prepolymerized chains of another monomer to form a long chain. The blocks of one polymer are not all of the same length. In graft polymerization short chains of a monomer, X, are grafted onto a long chain of Y, forming side chains. By the use of certain catalysts it is possible to orient the monomer groups in the copolymer chain in a predetermined regular order.

Condensation Polymerization. This is a reaction in which the bonding together of molecules results in the elimination of an element, such as nitrogen or hydrogen, or of a simple compound, such as hydrogen chloride, water, and sodium chloride. If both reactants are monofunctional, as in the reaction between ethyl alcohol and acetic acid, a relatively small molecule with no plastic properties is formed. If either the alcohol or acid is bifunctional and the other monofunctional, a larger molecule is produced. An example is diethyl phthalate which is a plasticizer but not a plastic polymer.

If both acid and alcohol are polyfunctional it is possible to produce a plastic polymer. For example, phthalic anhydride and glycerol (see Figure 1.1) react to form the alkyd resin which is made up of a long chain of repeating units. The remaining

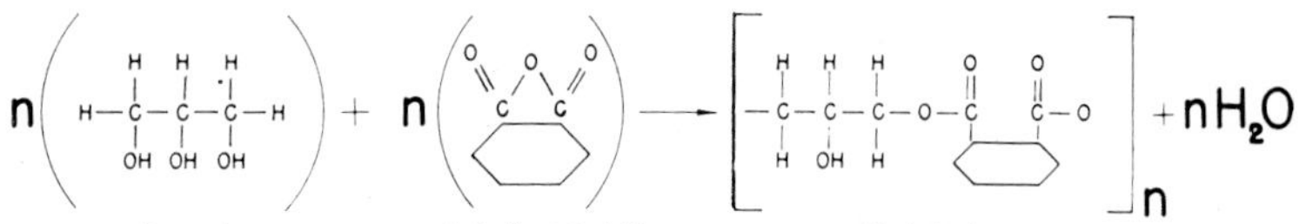

Fig. 1.1—Alkyd resin reaction.

OH group could then react with one hydrogen of another acid molecule to put on a side chain. This would then cross-link to another chain. Phenol, which has three reactive hydrogens, reacts with formaldehyde to form chains which cross-link and set up thermally. If one of the three reactive groups has been previously replaced by another radical as in para cresol, only a chain can be formed.

Mixed Polymerization. Some polymers result from a combination of addition polymerization and condensation. A relatively small amount of a very reactive compound may be copolymerized with one which ordinarily would form only a chain. This would result in chains which would then cross-link at the reactive points. Cross-linking of some thermal plastic chains can also be produced by X-ray or nuclear radiations.

Polymerization Reactions

Mechanism of the Reactions. While there are addition polymerization reactions, such as ethylene oxide in the presence of traces of water, which proceed stepwise, the typical addition polymerizations are chain reactions involving aliphatic double or triple bonds, as for example ethylene and its derivatives. The first step in these polymerizations is the activation of the monomer.

This activation requires, first of all, the addition of energy to break the double or triple bond. The amount of dissociation energy varies with the atoms making up the double or triple bonds. For example, a carbon to carbon double bond requires less energy than a carbon to oxygen double bond. The carbon to nitrogen double bond requires less and the triple bond more energy than the carbon to carbon bonds. Another factor affecting the ease of polymerization is the character of the atoms or radicals attached to the atoms forming the multiple bond. If one or both hydrogen atoms on *one* of the carbons of the double bond are replaced, the energy needed for activation becomes less. For example, ethylene is polymerized with conventional oxidizing catalysts with difficulty. If one hydrogen is replaced by a chlorine atom, vinyl chloride is formed. This frequently polymerizes spontaneously at ordinary temperatures when in contact with air. However, it is possible to substitute so heavily near the double bond that polymerization becomes more difficult, apparently because of steric hindrance. On the other hand, conjugation or cumulation of double bonds speeds up polymerization.

Initiators or Catalysts. Ordinarily an initiator of some type is required to start the reaction at a satisfactory rate. This is usually referred to as a "catalyst" although it usually reacts with the monomer. It is frequently used with an "activator." When a reducing activator is used with an oxidizing initiator the system is usually referred to as a *redox* system. The terms *catalyst, accelerator,* and *promoter* are often used interchangeably, as are *stabilizer, inhibitor,* and *retarder.*

The initiator or catalyst may be any chemical which decomposes under the polymerization conditions to give free radicals. Benzoyl peroxide is a common initiator in mass or solution polymerization. Hydrogen peroxide and potassium persulfate, since they are soluble in water, are commonly used in emulsion polymerization. The free radicals coming into contact with monomers react to give unpaired electrons on the carbons. These activated monomers come into contact with other monomers reacting to add on. The chains thus started continue to grow until terminated by collision with another growing chain, with the reaction vessel wall, with an impurity, or with a retarder or inhibitor. The latter are chemicals which react with the active chain ends to form nonreactive products. They are added to the system to control the chain lengths. Typical inhibitors are hydroquinone and phenol. The chain lengths are also affected by the temperature and the amount of the initiator used. With a large amount of initiator many chains are started, limiting the number of monomer molecules available for each chain.

Catalysts of the peroxide type produce polymers with a random arrangement with respect to the relative positions of the elements and radicals attached to the carbons in the main chains. By use of the so-called "stereospecific catalysts," of which the Ziegler type is an example, it is possible to direct the polymerization so that the relation of the substituents in the chain is oriented as desired. The Ziegler catalyst is the product of the reaction of a compound of a metal of groups IV to VIII in the periodic table, and an organometallic compound from groups I and III metals. An example is produced from aluminum triethyl and titanium tetrachloride. These catalysts are used in fluid-bed systems.

Another group of stereospecific catalysts are used in fixed-bed operations. One of these (Phillips Petroleum Company method) uses chromium oxide or a mixture of chromium oxide and strontium oxide supported on silica, alumina, silica-alumina, zirconium oxide, or thorium oxide bed. Two others (Standard Oil Company methods) use nickel or cobalt metal on charcoal and molybdenum oxide on alumina.

POLYMER STRUCTURES

Types of Molecular Structures

Straight Chain Polymers. The long linear chains formed by addition polymerization may be bundled together in a haphazard manner like straws in a stack. The atoms in the chains are held strongly together by primary valences. The bonds between the chains are secondary bonds or van der Waals forces resulting from the state of unbalance or unsaturation at the surface. These bonds, while they may vary considerably in strength, are normally much weaker than most primary bonds. They are effective only where the surfaces of adjacent chains are in contact. The interlacing of the chains may add some strength to the mass in a purely mechanical way.

This unoriented haphazard structure produces an amorphous material. Amorphous solids do not have sharp melting points. They break with a conchoidal fracture and tend to flow under stress. The magnitudes of the physical properties are not dependent upon direction.

If the chains are parallel, the polymer tends to be crystalline. The symmetry of the chains also affects the crystallinity. If the chains are perfectly uniform, they lie closely together, thus producing a stronger attraction and greater crystallinity than if they were irregular. Linear polyethylene (see A, Figure 1.2) is an example of this type of polymer. If part of the hydrogens are replaced by radicals and the substituted radicals are large, they may hold the chains somewhat apart, reducing the crystallinity. The arrangement of the radical, R, on the chain may be random (see B, Figure 1.2). This type of polymer, designated by Natta as "atactic," has a low degree of crystallinity. A regular arrangement such as C, or its inverted form called "isotactic," is

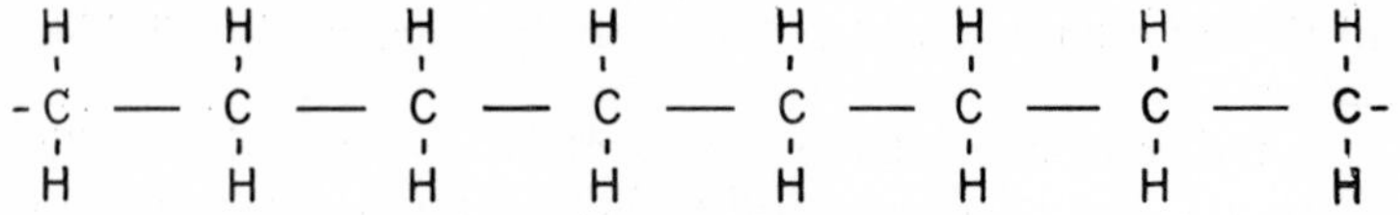

Fig. 1.2—Types of chain structure.

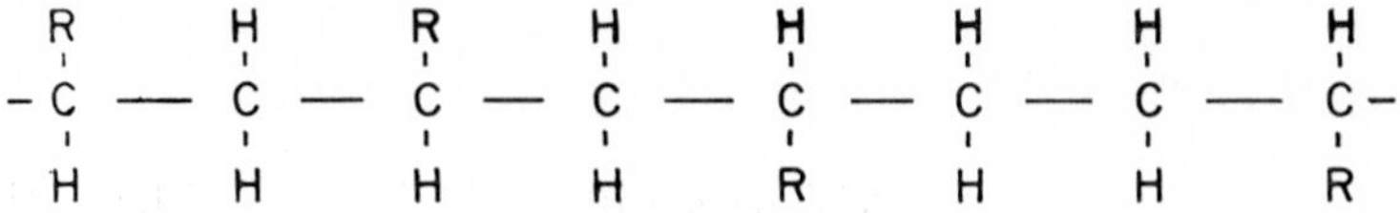

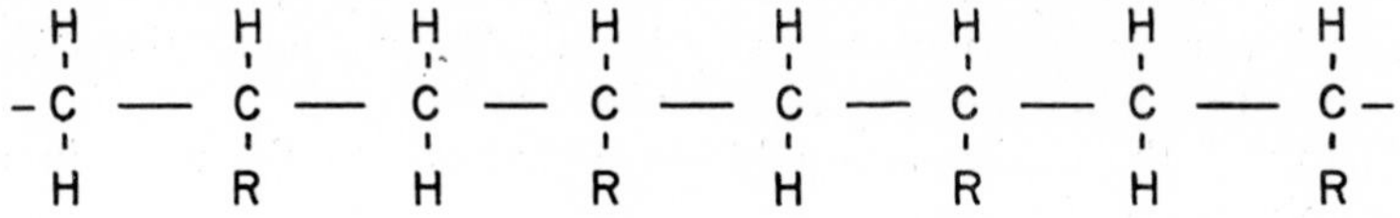

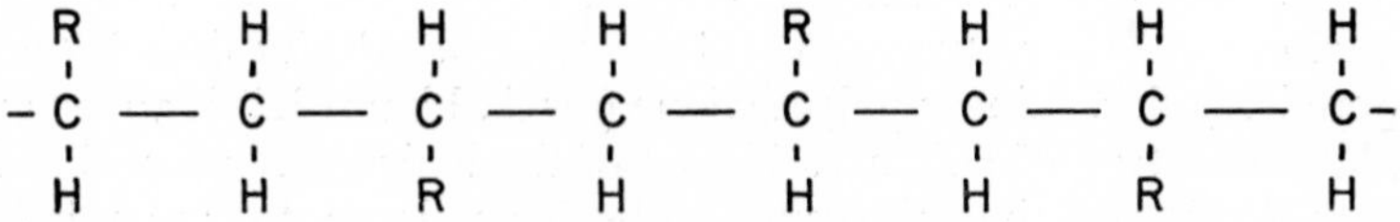

crystalline. It is also possible to have blocks of the two forms alternating in a regular order along the chain which would have considerable crystallinity. Another type of regular arrangement, shown in D, is designated "syndyotactic."

Sometimes the polymer is made up of both crystalline and amorphous areas. The crystallinity may vary with the conditions of solidification. It is frequently possible to orient the chains into parallel positions by stretching, thus changing from an amorphous to a crystalline product. The resin is then stronger along the direction of the chain length than in other directions.

These straight chain polymers, whether amorphous or crystalline, are thermoplastic; that is, they may be repeatedly melted by heat and solidified by cooling. They are, in general, soluble in certain organic solvents. The higher the degree of polymerization, the longer are the chains, the lower the solubility, and the higher the melting point. As crystallinity increases, penetration of solvent between the chains is less, resulting in lower solubility. The melting point is also higher since the molecules are held more tightly together.

When two monomers, A and B, are copolymerized, the resulting chains are usually not symmetrical. As the amount of component B is increased from zero to 50 per cent, the softening point lowers and the solubility increases. As the amount of B is further increased, the trend of the softening point and solubility is toward the values for B.

Branched Chain Polymers. Side chains may branch out from the main polymer chain. This usually reduces the symmetry of the molecule, resulting in lower crystallinity, strength, and melting point, and increased solubility. If the side chains cross-link between the parent chains with primary bonds, the effect is an increased melting point and decreased solubility.

Moderately Cross-linked Polymers. Cross-linkage by primary bonds between chains holds the chains more rigidly together and reduces lengthwise slippage. Soft vulcanized rubber, a typical elastomer, which has from five to ten sulfur cross-links for each hundred linear bonds, is no longer thermoplastic. Stretching allows a limited amount of reversible slippage of the chains by each other. Solvents can still penetrate between the chains. In some of these the rubber is soluble. Others merely cause swelling.

Highly Cross-linked Polymers. These polymers are ordinarily produced by condensation or by a combination of addition polymerization and condensation. In the intermediate stage before converting to the final form they are usually chain compounds. The condensation, or cross-linking, reaction occurs upon heating. The finished product is infusible and insoluble in most solvents. This type of resin is known as thermosetting.

The number of cross-links per 100 linear bonds is known as the netting index. The netting index of hard rubber is from 10 to 20. A thermosetting resin such as a phenolic will have a netting index of about 50.

Molecular Weights

Types of Molecular Weights. If a given polymer were made up of identical molecules, the molecular weight, M_n, could be determined by dividing the mass by the number of molecules in the mass. In polymerization processes all molecules do not attain the same size, that is, the same degree of polymerization or number of component units. The molecular weights of polymers are therefore expressed as average molecular weights. The average of M_n would be the *number-average* molecular weight, $\overline{M}_n$.

The number-average molecular weight does not define the various sizes of molecules and numbers of each size. Thus two polymers might have identical number-average molecular weights and yet have quite different molecular weight distribution and consequently different properties. If we could determine the number of molecules of each different molecular weight in a polymer it would be possible to determine the mass of each molecular weight group. Dividing the sum of these masses by the total number of molecules would secure the weight, or mass, average molecular weight, $\overline{M}_w$.

Molecular Weight Determinations. The conventional methods of molecular weight determination are difficult to apply accurately to polymers with molecular weights above 10,000. Most of the plastic polymers run from around 20,000 to several hundred thousand or even over a million. The determinations may be made on the original polymer or upon fractions segregated over certain molecular weight ranges.

Freezing-point lowering, vapor-pressure lowering, and boiling-point elevation are of little value above a molecular weight of 5,000 or possibly up to 10,000. The results are number-average molecular weights.

Osmotic pressure determinations are used on the determination of number-average molecular weights up to about 500,000 on polymers which are soluble. Since this depends upon the measurement of the pressure developed when a solution is separated from the solvent by a semipermeable membrane, the sizes of molecules which can be handled successfully depend partly upon the membrane used. Static determination of the pressure is made with a calibrated capillary tube. In the dynamic method the osmotic pressure is counterbalanced to zero flow by an applied pressure of known amount. The molecular weight for low molecular weight polymers can be determined by the van't Hoff equation, $M = RT/\pi/C$, where π/C is a constant. For the high polymers π/C may be determined by extrapolation of experimental data.

Ultracentrifuge methods, originally used on proteins, have had some application to plastic polymers. They are based on the principle that particles suspended in a liquid arrange themselves under the influence of gravity so that the vertical distribution is proportional to their weights. However, even large polymer molecules will not show sufficient distribution under direct gravity to be of value for measurement. By the use of an ultracentrifuge at about one million times gravity it is possible to check the molecular range. Results are expressed as $\overline{M}_z$.

Viscosity determinations of polymer solutions or liquid polymers are probably the most commonly used methods for molecular weight estimations. The viscosity relations of polymer solution⁻ are complicated by attraction between the solvent and solute and the variation in molecular size and configuration. The basic relation is that the viscosity is equal to a constant specific for the polymer and solvent times the average molecular weight times the monomer molecular weight. Since this holds only for the smaller polymers it has been modified empirically for various applications. Empirical viscosity methods are used for control purposes in the manufacture of various plastic polymers.

One type of viscosity determination is known as *melt index*. This is the number of grams of molten polymer which will flow through a standard orifice at a standard pressure and temperature in a given time. The melt index is roughly a logarithmic function of the molecular weight. It is used mainly as a plant control method although it gives a rough indication of molecular weight.

Light scattering, which results when a beam of light is

passed through a solution of a polymer, can theoretically be used to determine the weight-average molecular weight. The light scattering is the additive result of the scattering produced by the solvent and the polymer molecules. This method is applicable to molecular weights above 10,000, particularly in low concentrations. Special tubidimeters have been developed for the measurement. Chemical methods of determining molecular weight from the number of end-groups on the molecules have been used. These methods vary with the different end-groups such as carboxyl, hydroxyl, and acetate units. Reagents reacting only with the groups being determined must be used. The structure of the polymer must be known and it must be free of reactive impurities. Since the number of molecules are determined, the results are in number-average molecular weights. Functional groups may be also introduced into the molecule during polymerization to act as tracers.

Molecular Weight Distribution. Distribution curves plotted between either molecular weights or degree of polymerization and weight per cents or number of molecules are frequently used to characterize specific samples of a polymer. The necessary data in some cases can be determined by ultracentrifuge or light-scattering techniques. Polymers may also be fractionated by gradual addition of a precipitant to a solution of the polymer, by changing the temperature of the solution thus changing the solubilities of various fractions, and by evaporating part of the solvent to precipitate certain fractions. The molecular weights of the various fractions are determined by one of the methods previously described.

POLYMER MODIFICATION

The modification of polymers to change their characteristics is of considerable importance in the plastics industry. A few general principles and examples of what can be done will be given to indicate possibilities.

Chemical Modification

The substitution of chlorine or fluorine in polymers tends to increase the melting point and resistance to chemicals. For example, substitution of the hydrogens in polyethylene by fluorine produces polytetrafluoroethylene, a polymer with an increase in continuous heat resistance from 220° F to 500° F. It also has outstanding chemical resistance. Polydichlorostyrene has a heat distortion temperature of about 265° F compared with poly-styrene at 195° F.

Increasing the hydrocarbon content of a plastic polymer tends to decrease moisture absorption. For example, cellulose acetate-butyrate has lower moisture absorption than cellulose acetate. Polyethylene has lower moisture absorption than poly-vinyl chloride. Degree of polymerization may affect moisture absorption or water solubility. A given polymer with a low molecular weight may be water soluble but insoluble if pro-duced with a high enough molecular weight.

Cross-linking by the addition of another compound may be useful if there are suitable unsaturated positions in the chains where the cross-links may be attached. For example, cross-link-ing of raw rubber with sulfur (vulcanization) may produce the well-known soft rubber. Here the linkages are not numerous enough to prevent a limited movement of the molecular chains over one another and thus allow stretching, but are strong enough to return the rubber to its original position when the stress is removed. It is also possible to cross-link at more points and produce a rigid material, hard rubber.

If there are no suitable unsaturated positions in the thermo-plastic polymer chains it may be possible to introduce some by copolymerizing with an unsaturated compound. In butyl rubber just enough isoprene is copolymerized with the isobutylene to provide the proper amount of cross-linking without leaving any residual unsaturated points at which oxidation or other unde-sirable reactions can occur later. Certain cross-linking chemicals may increase heat and flame resistance. Examples are triallyl cyanurate and certain phosphonate compounds used with poly-esters.

Structural control by specific catalysts of the Ziegler type is used in the polyolefins to produce the straight chain crystalline molecules rather than the weaker lower melting point branched chain molecules.

Plasticizers

Plasticizers are high-boiling liquids, usually organic esters, or low-melting solids, such as camphor, which are added to plastic polymers to modify their physical properties. Originally, as the name implies, their primary purpose was to render the resin mixture more plastic. They are frequently necessary in thermo-plastic resins to produce improved flexibility in such end products as vinyl raincoats or hose. Sometimes they are necessary in both thermoplastic and thermosetting plastics to promote better molding properties, such as improved flow, during fabrication.

Solvents may also be used in the initial compounding of thermoplastic molding compounds or in the formulation of surface-coating products such as lacquers. Since the solvents eventually vaporize, the properties imparted by them are temporary. Plasticizers, on the other hand, if properly selected, impart permanent properties to the product. Primary plasticizers can be used alone without incompatibility. Secondary or extender plasticizers must be used with a primary plasticizer. Copolymerization is sometimes referred to as an internal plasticization since the introduction of the second monomer even in small quantities may make considerable change in the properties of the finished compound. In addition to improving flexibility, plasticizers may modify one or more of the following characteristics: elongation, tensile strength, toughness, softening point, melting point, heat-sealing characteristics, resistance to flow under pressure, flame resistance, electrical properties, water absorption, oil resistance, abrasion resistance, and toxicity.

Mechanism of Plasticizer Action. The attractive forces between thermoplastic resin chains are, as we have seen, secondary bonds or van der Waals forces. Both solvents and plasticizers can enter between the chains, thus reducing the forces holding them together. This modifies the properties of the resin, depending both upon the magnitude of the original forces holding the chains together and the polarity of the plasticizer. Polar plasticizers with assistance of heat and sometimes solvents are needed to plasticize such strongly bonded chains as those in cellulose acetate and polyvinyl chloride. The polymer-polymer bonds may be replaced by polymer-plasticizer bonds, thus not only producing flexibility but additional toughness. It is possible in some cases to add sufficient plasticizer so it forms a continuous phase in which the chains move. Such a plasticized mass may be a soft gel.

Choosing a Plasticizer. The following need to be considered in selecting a plasticizer. *Low volatility* is important. If the plasticizer evaporates in a short time under use, the product suffers deterioration in desirable properties. *Toxic plasticizers* should be avoided. Some plasticizers produce dermatitis upon contact with the skin. Others may produce internal toxic effects from contact with food or drink. *Solubility in solvents* with which the finished product comes into contact is undesirable. *Stability* to such factors as moisture and sunlight is important. *Migration* into other materials, particularly other plastics with which the product comes into contact, is undesirable. *Corrosive* effect upon metals in contact with it may be a serious problem.

Typical Plasticizers. Several phthalates, including the dimethyl, dibutyl, dioctyl, butyl benzyl, and diisodecyl, are used in thermoplastic materials to improve flexibility. Triphenyl and tricresol phosphates are used for their flame-proofing properties. Among other plasticizers are camphor, vegetable oils, adipates, sulfonamides, phthalyl glycollates, ortho nitrobiphenyl, butyl oleate, and methoxy ethyl stearate.

Irradiation

Ultraviolet Light. All rays with wave lengths below those of visible light probably have some effect on all plastic polymers. Ultraviolet rays activate certain polymerization reactions. The ultraviolet rays in sunlight are sufficient to cause deterioration of some plastic materials, such as rubber and polyethylene, unless suitable stabilizers are added.

Ionizing Irradiation. While it has been shown that X-rays have definite effects on plastic polymers, these effects are small compared with the action of the gamma rays and neutrons produced by pile radiation. Two general effects have been observed: cross-linkage of certain thermoplastic polymers and degradation of certain other polymers by scission. While it is possible to classify polymers into the two above categories, the classification is not rigid. Both cross-linking and scission may go on simultaneously, the predominating reaction determining the final result. The presence of oxygen during irradiation tends to accelerate breakdown of some polymers but apparently has little if any effect on others.

Cross-linkage is believed to occur after the removal of a hydrogen atom on each of two adjacent chains. Polyethylene, polystyrene, polyacrylic esters, nylon, natural rubber, neoprene, and butadiene-styrene rubber are among the polymers which may be cross-linked. For example, rubber can be vulcanized by irradiation. Unless the irradiation is controlled, the rubber may become stiff, losing its extensibility. Eventually it cross-links sufficiently to be hard rubber.

It has been found that a monomer, such as methyl methacrylate, can be impregnated into wood and polymerized by radiation, preferably gamma rays from cobalt 60.

Degradation From Irradiation. All plastic materials eventually suffer some degradation even though the initial effect may be to cross-link. Certain additives may reduce the effect of radiation deterioration. For example, certain organic compounds, such as N, N'-cyclohexylphenyl-p-phenylene diamine, added to rubbers considerably increase their resistance to ionizing radiation. Mineral fillers increase the resistance of phenolics and polyesters. Polyethylene and polystyrene have good resistance. The addition of carbon black to polyethylene increases its resistance to ultraviolet greatly and to ionizing irradiation somewhat.

CHAPTER 2

Production of Plastics Materials

GENERAL PRODUCTION METHODS

Methods of Carrying Out Addition Polymerization

Mass or Bulk Polymerization. The simplest method of polymerizing a monomer is allowing it to stand at the proper temperature with a catalyst if necessary until polymerization is complete. Theoretically, this can be done with a solid, liquid, or gaseous monomer. Practically, most polymerizations are carried out in the liquid phase. The polymer may be either soluble or insoluble in the monomer. In the first case the mass becomes progressively more viscous until polymerization is complete. In the second case the polymer tends to settle out and may be separated from any remaining monomer.

One of the disadvantages of mass polymerization is the difficulty in removing the heat of reaction which is usually in the range of 10,000 to 20,000 calories per mole. As the reaction mass becomes more viscous it becomes increasingly difficult to stir, making heat dissipation difficult. The molecular weight distribution is difficult to control. Bulk polymerization frequently works out well in production of small castings. Continuous polymerization in a tower has been used for the production of molding powders. Another method is to spray a polymer-monomer mixture into hot inert gas. Mass polymerization in gaseous phase is carried out under pressure with gases such as ethylene and other unsaturated gases, both with and without catalysts.

Solution Polymerization. The monomer may be polymerized in a solution of a suitable solvent. This simplifies heat removal since the solution is less viscous than the molten resin. Heat can also be removed by refluxing the solvent.

Certain disadvantages also are inherent. The polymerization temperature is limited by the boiling point of the solvent, thus slowing down the reaction. The average molecular weight is usually lower. Separation and recovery of the solvent may be expensive. If the solvent is flammable or toxic, safety problems must be solved. If a suitable liquid in which the monomer is soluble but the polymer is insoluble is used, the separation of the polymer is simplified.

Emulsion Polymerization. If the monomer can be polymerized in a water emulsion the heat removal problem is simple and some of the disadvantages of solvent polymerization can be avoided. The reaction rate is greater and the molecular weight of the polymer is usually higher than with bulk or solution polymerization. Water, even if deionized as sometimes is necessary, is cheaper than solvent and involves no recovery problems. However, the cost of the emulsifying agent and the separation and drying of the finely divided polymer add to the expense.

Suspension Polymerization. Suspension, or pearl, polymerization is carried out with globules maintained in suspension by proper agitation without an emulsifying agent. The monomer droplets may be as large as 10 microns. The final beads formed by coalescence are regulated in size by the agitation and by the use of suitable suspension stabilizers. These stabilizers may be organic materials such as gelatin or starch, or inorganic materials as talc or diatomaceous earth. The pearls are readily separated from the water by screening and are washed free of stabilizer by water.

Methods of Carrying Out Condensation Polymerization

Condensation polymerization is commonly a mass or bulk polymerization. Provision must be made for removal of the by-product material, such as water. Since these reactions ordinarily produce thermosetting compounds the reaction must be stopped before completion to produce a thermoplastic product. The reaction may be stopped by cooling, in which case it is completed by heating in the final molding operation. Another method is to add only part of one reactant in the preliminary reaction. The thermoplastic material is then mixed with the final part of the one reactant and molded to complete the polymerization. When the reactants and the initial thermoplastic polymer are water soluble, as for example in urea-formaldehyde polymers, the first part of the polymerization may be carried out in water solution. The water is removed by evaporation and the polymerization completed in the mold. Alkyd resins are sometimes made by reacting the constituents in a solvent. The solvent is continuously refluxed back into the reactor. Condensation polymerization is also possible in the gas phase.

POLYMER PROCESSING EQUIPMENT

Resin Reactors

The typical plastic resin reactor (see Figures 2.1 and 2.1A) used for mass and solution polymerization is a vertical closed cylindrical vessel with a top-driven agitator. Provision is made for charging and discharging through suitable valves, for cleaning the interior, for sampling, and for instrumentation. While the proportions vary, the height is usually greater than the diameter. The amount of head room allowed above the normal liquid level varies with the amount of foaming and may exceed the diameter. Bottoms are usually rounded. Capacities vary with need, but are commonly between 1,000 and 2,000 gallons. They are designed to handle pressures from vacuum to 15 psi internal pressure for phenolic processing up to 20,000 psi for high pressure polyethylene processing. The standards normally followed are those of the A.S.M.E. code for unfired pressure vessels.

Materials of construction vary with materials being processed but include carbon steel, copper, aluminum, stainless steels, "Monel," "Inconel," nickel, and nickel and stainless clad. Probably the most common material is 18-8 chrome nickel stainless steel. *Heating* systems include direct fired, hot water, steam, hot oil, electrical resistance, "Dowtherm," and "Electro-vapor." For temperatures below 350° F., as for example in the production of phenolics, ureas, and melamines, steam is commonly used. Other media are hot water, hot oil, and liquid "Dowtherm." The heat is applied through a jacket or internal coils. One advantage of these systems is that the jacket or coils can also be used for

Fig. 2.1—Plastic resin reactor. (Courtesy Blaw-Knox Company)

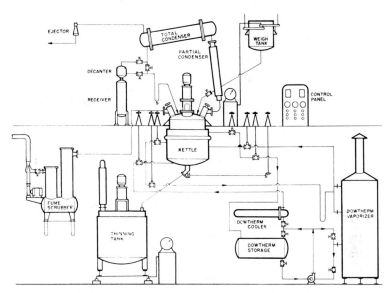

Fig. 2.1A—Alkyd resin unit—an added possibility which is available.
(Courtesy Blaw-Knox Company)

circulating a cooling fluid. For higher temperatures direct firing, electrical resistance, "Dowtherm," and "Electro-vapor" heating may be used. Electrical resistance heating is readily controlled but may be expensive and provides no jacket or coils for cooling. "Dowtherm" under pressure is popular for high temperatures, being used much like steam but with lower pressures for a given temperature. "Electro-vapor" uses "Dowtherm" heated by electric immersion resistance-heating elements and may be used up to 650° F. The "Dowtherm" is boiled in the lower expanded portion of the reactor jacket. The vapors rise up the jacket, heating the kettle. Direct heating by gas or oil flame was one of the earliest methods used for high temperature varnish resin processing and for polymerization. One disadvantage was the danger of overheating in spots. A newer development of this is radiant heating, using gas or oil as the fuel. The furnace consists essentially of a light, highly insulated, stainless steel shell. This is heated by tangentially mounted burners. The reactor is suspended in the furnace and is heated mainly by the radiant heat from the inside black surface of the furnace. This type of heating is said to be easily controlled and low in cost.

Agitator types (see Figure 2.2) and speeds vary with the material being processed. For mass polymerizations where viscosities of 50,000 poises or more may be involved, the close-

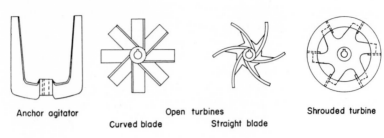

Anchor agitator   Open turbines   Shrouded turbine

Curved blade   Straight blade

Fig. 2.2—Agitator types.

fitting *anchor* or *horseshoe* agitator is popular since it is effective in preventing a thick buildup on the reactor wall. Since in most cases the resin film heat transfer coefficient is approximately equal to the overall coefficient, it is desirable to keep the wall well scraped. The coefficient for a well-scraped wall may be several times an unscraped one. Sometimes the horseshoe agitators are equipped with spring-loaded blades which give a very effective scraping action. Swirl stoppers or baffles may be necessary to prevent the contents of the reactor rotating without effective mixing. When provided with a variable speed drive, horseshoe mixers may be used for mixing materials of different viscosities. Cross arms are sometimes added to produce more effective agitation. *Paddle* agitators are also used for mixing high viscosity materials.

    Turbine mixers which function somewhat like a centrifugal pump are excellent for some materials. *Shrouded turbines* are effective at moderate speeds in viscosities up to about 5,000 centipoises. *Open* turbines are suitable up to about 50,000 centipoises. *Propeller agitators* are suitable only for low viscosity materials, such as below 20,000 centipoises. Agitators are direct-driven from the top of the reactors by electric motors. Multi-speed motors may be used to provide for more efficient utilization of the reactors with different polymer systems. Explosion-proof motors and wiring are frequently desirable.

    Temperature control, because of the high heats of polymerization, is necessary to prevent damage to thermoplastic resins and premature hardening of thermosetting resins. In reactors heated by fluids in coils or jackets, the temperature may be controlled in part by circulating water or other cooling media through the heating channels. Reactors heated by electricity or radiant heat may have internal cooling coils. Where condensable vapor results from the reaction, *reflux condensers* are useful for heat removal. These are ordinarily water-cooled tube and shell condensers, sitting vertically on the tops of the reactors. Inclined condensers for the condensation and removal of volatiles are also frequently provided, particularly for handling condensation reactions.

Instrumentation ordinarily includes at least pressure gauges and thermometers either or both of which may be of the continuous recording type. *Sight glasses* and *sampling* tubes or valves are frequently needed. Provision for adding an inert gas to blanket the reaction may be necessary. The reactor should be equipped with a *safety pressure relief valve.* In addition *frangible discs* are frequently added, especially if there is a possibility of malfunctioning of the safety valve as the result of corrosion by vapors.

Emulsion Reactors

Reactors used for emulsion and suspension polymerization are similar to those used for resins. Those used for synthetic rubber are jacketed for cooling and usually made of stainless steel or glass-lined steel. Agitators are commonly stainless steel paddle, turbine, or propeller types, operating at high speeds. Horizontal rotating nickel- or glass-lined reactors about three feet in diameter have been used for emulsion polymerization of polyvinyl chloride.

Continuous Reactors

Continuous emulsion polymerization of synthetic rubber has been carried out in a series of kettles through which the reacting matter was pumped. Continuous reactors for polymerizing vinyl chloride in emulsion form used in Germany were equipped with agitators in the top one-third. The polymer settled to the bottom, from which it was removed continuously. Polystyrene has been polymerized in mass in 19-foot high towers equipped with jackets for heating and cooling. It was removed from the bottom onto a stainless steel cooling belt. Belts have been used in Germany for continuous polymerization of isobutylene in ethylene solution. The belts were of stainless steel 12 inches wide and formed into a 3-inch deep trough by idler pulleys set at an angle under both sides. The belt operated through a 30-inch in diameter steel shell. A similar belt of rayon was used for polymerizing polyvinyl isobutyl ether.

Bulk Polymerizers

Some polymers are polymerized in a mass without agitation after the constituents have been combined in a suitable mixer. Cast phenolics are polymerized in molds to produce the desired shape. Frequently these are thin lead shells formed by dipping steel forms, such as rods, into molten lead. The lead is stripped off the finished plastic piece and remelted.

Polystyrene is sometimes produced by polymerizing the monomer previously mixed with the catalyst in a suitable mold. Originally, because of the difficulty in controlling the temperature, only thin sheets were made in this way. Now blocks as large as 24 inches by 50 inches by 5 inches are polymerized in frames between water-cooled platens in a modified plate and frame filter press. The plates and frame are made of cast aluminum.

Miscellaneous Reactors

Cellulose esters such as the acetate are commonly produced by reacting the cellulose with the acid or anhydride in a Werner-Pfleiderer type of mixer equipped with a water jacket. These were originally made of bronze but stainless steel is now used. Rotating copper drums about 11.5 feet in diameter have also been used in Germany. Originally cellulose nitrate was produced in conical-bottomed stoneware crocks with a false bottom. The more recent nitrating vessels are covered cast-iron or steel kettles equipped with agitators and cooling jackets. Viscose for cellophane and rayon is produced in a rotating "barratte," followed by a mixer and ripening tanks. Other types of reaction vessels are also used.

Dewatering Equipment

If the plastic material is polymerized by either the emulsion or suspension technique it must be separated from a considerable amount of water. The relatively coarse suspension particles are more readily filtered from the water and dried than are the finer emulsion particles. *Spray drying* has been used, for example, on polyvinyl chloride emulsions. It has two obvious disadvantages: the high cost of evaporating large amounts of water and the retention of the emulsifying ingredients in the final product. *Filtering* on a rotary filter or a vibrating dewatering screen is frequently used ahead of different types of drying equipment.

Centrifuging may be used also to remove the bulk of the water before drying. When desired, the polymer may be washed on the filter or centrifugal to remove much of the emulsifier before drying. *Drying* may be done in a rotary dryer or in a tunnel dryer on continuous belts.

Extrusion drying is a new development used on synthetic rubber. After passing over a dewatering screen the rubber "crumbs" are mixed with wash water in a leaching tank and then dewatered on another screen. From here they go to the extruder. The conical screw in the extruder compresses the rubber as it moves through, squeezing out about 80 per cent of the water. The rubber is carried from the end of the extruder proper through a heated section under a 27 mm Hg vacuum by a screw conveyor. At the end of the screw conveyor it is forced through a perforated plate and cut into pellets by a revolving knife. Because much of the water is squeezed out rather than evaporated, the ash content of the rubber is low. A unit in operation is said to have a capacity of 8,000 pounds per hour. It dries the rubber to 0.5 per cent moisture in a total time of 5 to 10 minutes.

A novel dryer used for removing both water and solvent from plastic polymers is the *conical vacuum dryer* (see Figure 2.3). It is made up of two vertical cones connected by a circular

Fig. 2.3—"Conaform" dryer. (Courtesy Patterson Industries)

band or short cylindrical section. The unit rotates about a horizontal shaft. It is jacketed over both cones and the center section for the circulation of hot water or steam for heating. The unit operates under reduced pressure. As the dryer rotates, the material is constantly changing position, coming into contact with the hot surfaces and also becoming a part of the surface from which evaporation occurs. Units as large as 9 feet in diameter are in use.

Desolventizing

Solvent is removed in a reactor by vaporizing and condensing. If water is present it may be necessary to run the condensate into a tank or solvent water separator to separate the two by gravity. Solvent may be removed like water in a spray dryer or in the vacuum dryer previously described. Tight equipment is necessary to avoid solvent vapor leakage because of both cost and safety considerations.

Pumps

Practically all types of pumps are used in various phases of plastic polymer production. Many of the liquid monomers present no unusual problems, being handled by centrifugal and rotary gear pumps. Polymer emulsions can be handled by centrifugal pumps if they do not break the emulsion by the mechanical agitation. Diaphragm pumps were originally used for handling synthetic rubber latex to avoid any formation of "prefloc." Because of excessive diaphragm breakage, open impeller type centrifugal pumps were tried. They are very satisfactory.

Plastic polymers and polymer solutions are usually too high in viscosity to be handled by the common types of pumps. Materials with 5,000,000 SSU can be pumped in screw pumps (see Figure 2.4). These contain two pairs of meshed rotor screws operating in a casing. The product being pumped enters at both ends and is carried by the rotor screws to the center discharge. Another pump suitable for these high viscosity materials is the "Moyno" pump. A special rotor produces a positive nonpulsating pushing action within a casing giving a uniform flow. A rotary gear pump using herringbone pumping rotors can also be used for pumping high viscosity materials. Materials of construction of the pumps used vary with the material being handled. Stainless steels are commonly used.

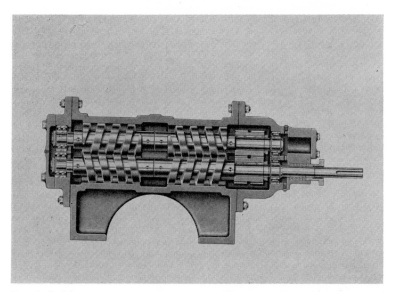

Fig. 2.4—Screw or gear type pump. (Courtesy Sier-Bath Gear and Pump Company)

Tanks

Tanks are used to handle raw materials—for "aging" (completing a reaction started in a reactor) polymers, for dissolving resins, and for simple storage of finished products. Materials of construction vary with the material being handled. Typical materials are ordinary steel, stainless steel, aluminum, and wood. Agitators of the various types and sometimes heating coils used in reactors are used in tanks for such operations as aging and dissolving. Provision may also be made for maintaining an inert gas, such as nitrogen, over monomers adversely affected by contact with air.

PRODUCTION OF MOLDING MATERIAL

Production of a molding material from the polymer may simply involve grinding or pelleting. More often the polymer must be intimately mixed with plasticizers, fillers, extenders, colorants, stabilizers, lubricants, and catalyzers. This may involve either dry or molten blending. Additives and equipment will be described briefly.

Additive Materials

Fillers. Fillers are used primarily to improve the properties of the finished product. They frequently reduce brittleness and increase impact strength. Usually they are cheaper than the resin and therefore lower the overall cost. In most cases the effect is mechanical, the fillers not reacting with the resin. *Wood flour,* made mainly by grinding fir, pine, spruce, basswood, and cottonwood to approximately 100-mesh size, is the common general purpose low-cost filler used in phenolics. Blends of it and the phenolic resin mold readily, resulting in products of good mechanical strength and appearance. *Alpha cellulose,* cotton linters, or highly purified wood pulp are used to a lesser extent mainly in urea and melamine plastics. *Cotton flock,* short cotton textile clippings, is used where greater impact strength is desired. *Sisal fibers* and *cords* are used for still greater impact strength. *Walnut shell* powder produces a molded product with a smoother finish but less strength than wood flour. *Chopped paper* is also used as a filler.

Asbestos floats or fiber products are used in phenolic products when high heat resistance is desired. Asbestos-filled products also have low water absorption and good chemical resistance. *Mica*-filled phenolics have excellent electrical properties and good heat resistance. *Diatomaceous earth* is used for its heat and chemical resistance properties. It has somewhat poorer electrical and heat resistance properties than mica and has considerable abrasive action on the molds. *Amorphous silica, calcium carbonate,* and *clays,* such as kaolin, are used to improve flow and increase opacity particularly in glass-reinforced polyester products. *Glass fibers* are used in short lengths as fillers in polyester, epoxy, and phenolic molding compounds. In the form of longer filaments in rovings and mats they are used as reinforcements for polyester and epoxy resins, usually in relatively large pieces. A variety of inorganic powders are used as fillers in potting compounds.

Carbon black is used as a reinforcing filler in rubber and to a more limited extent as a filler and colorant in other plastic materials. *Graphite* is used in phenolic plastics because of its chemical resistance or because of its lubricating properties. *Synthetic fibers* such as nylon, acrylics, and polyesters are used as fillers and reinforcements to produce improvements in electrical, chemical, and mechanical properties. They are used in the form of fibers, batts, and woven fabrics.

Extenders. Certain materials are sometimes added to lower the cost without very noticeable changes in properties of the product. They are frequently referred to as *extenders,* although they are sometimes classified as fillers. Various types of *lignin* may be added to phenolic resins and to rubber. Protein material such as soybean oil meal may also be added to phenolic resins. This probably reacts with the aldehyde to give a mixture of phenolic and protein plastic materials. *Rosin* is sometimes used as an extender.

Colorants. These may be *dyes, inorganic pigments,* or *organic pigments.* In choosing colorants for a specific plastic, consideration must be given to other items than the color of the colorant. Some colorants are unstable under acid conditions, others under alkaline conditions. Others are sensitive to oxidation or reduction. These conditions may result not only from the chemical nature of the reacting monomers or the polymers but from the presence of catalysts or from degradation products produced in processing or in use. The finished product may in normal use come into contact with chemicals. Ultraviolet light causes a breakdown of some colorants unless a suitable stabilizer is present. Excessive heat in processing, molding, or use may break down the colorant.

Dyes are organic chemical compounds used mainly for coloring transparent and translucent materials. They are available in a wide range of colors, have excellent brightness, low plasticizer absorption, and low dielectric strength. Dyes have poor resistance to light and heat. They have a tendency to bleed. *Inorganic pigments* are largely metallic oxides which are much superior to organic colorants in weather resistance, in stability to light and heat, and in resistance to bleeding. They are available in a limited range of colors which are somewhat weak tinctorially. They are used in opaque plastics such as phenolics.

Organic pigments, in general, have a limited solubility, high oil and plasticizer absorption, and excellent dielectric properties. They are available in many colors which have good brightness and high coloring strength. They have better resistance to light and heat than dyes, but poorer resistance than inorganic pigments. They are used in practically all plastics except silicones. Colorants may be added either by the manufacturer or the molder. A concentrated mixture of the colorant with a resin may be added instead of the pure colorant.

Stabilizers. Some plastic resins deteriorate under the influence of heat and light during molding, some weather in the final molded form. This deterioration may be the result of oxidation from either residual or atmospheric oxidation or of partial breakdown of the polymer itself. Various lead and organic tin compounds as well as barium, cadmium, and zinc soaps have been used as stabilizers. Ultraviolet stabilizers include carbon black in polyethylene and benzophenone in vinyls.

Lubricants and Mold Release Agents. Various lubricants such as oils, waxes, and metallic soaps have been applied to molds to facilitate removal of molded pieces. Zinc, magnesium, lead, and aluminum stearates in the powdered form are frequently mixed with the other ingredients of the molding powder, usually in amounts of less than 1 per cent. They promote a more even flow of the plastic in the mold and also act as mold release agents. Calcium and lead stearate are used with vinyl compounds since they neutralize any free acid which may be present during molding. Dimethyl polysiloxane is used as a mold release agent by applying directly to the mold surfaces.

Equipment

Grinders and Cutters. Various types of standard grinders are used, depending on the product and size reduction desired. Large brittle lumps such as phenolic resin may be first crushed in a *jaw crusher* or a modified hammer mill. *Hammer mills* and *disc grinders* are used at the next stage. *Ball mills* are frequently used for the final stage to produce a powdered material. *Impact grinders* are also used. *Rotary cutters,* rotary knives acting against stationary blades, are frequently used for granulating thermoplastic materials to proper molding size. Thermoplastic materials are also frequently pelletized by *extruding* into small rods which are cut into short lengths.

Dry Blenders. Mixing of dry powdered or granular material can be done in a *ribbon,* or double-helical, *mixer.* The mixing is by two helical screws, one right hand and one left hand. *Tumble-type blenders* of the drum type with horizontal rotation are sometimes used, but unless properly baffled tend to produce rather poor mixing.

Double cone blenders, consisting of two cones joined by a

short cylinder, are common for dry blending. The blender rotates about a shaft perpendicular to the cylinder, producing a constantly changing flow cross section which results in good mixing. The *twin-shell* tumbler or rotor blender is formed from a cylinder cut at an angle and welded together to form a V-shape. Rotation is about a shaft crosswise of the V. Like the double cone blender the twin-shell blender produces a changing cross section. It is made with working capacities up to 1,700 cubic feet.

Hot Blenders. Rolls are frequently used for blending molten plastic mixtures. In the simplest form they consist of two hollow rolls set horizontally with provision for heating with steam or hot water and cooling by water. They travel towards each other at the same speed or with a small differential in speed. The sheet formed on the rolls is removed mechanically or manually and usually fed back to the rolls or to another set. *Sigma-blade* type "dough mixers" with heavy revolving arms are also used.

Banbury mixers (see Figure 2.5) consist of two rotors which move towards each other in a heavy jacketed casing. The temperature of the mixing operation is controlled by the steam or water supplied to the jacket. The material is held down in the mixing cavity by a floating weight. Banbury mixers are useful for mixing in fillers and colorants, for blending different batches, and for plasticizing resins and elastomers.

SAFETY PROBLEMS

The safety problems in polymer production are those common to chemical manufacture and processing, and the same general precautions and safety measures apply. All pressure vessels should comply with the A.S.M.E. code and be equipped with adequate pressure relief devices such as frangible discs or safety "pop-off" valves. Since chemicals are being handled, pressure equipment should be inspected regularly for corrosion which might weaken it.

Fire Hazards

Except for cellulose nitrate, the *solvents, diluents,* and *plasticizers* present greater fire hazards than the plastics. Cellulose nitrate presents a great fire hazard and must be handled

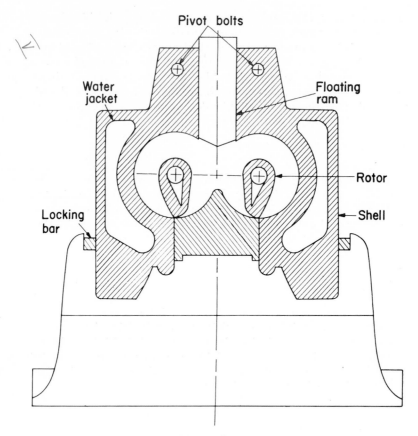

Fig. 2.5—Banbury mixer.

with special precautions. Solvents which are relatively safe at room temperatures may become heavy fire hazards when hot. The mixing of hot resins and solvents may produce flammable fumes which form explosive mixtures with air. Plasticizers vary widely in fire hazard. Volatile flammable materials must be kept in tight tanks or mixers. Blanketing with an inert gas such as carbon dioxide or nitrogen sometimes is desirable.

Dusts such as those from the plastics or from fillers such as wood flour may form explosive mixtures with air. Good housekeeping tends to lower this hazard. In all areas where flammable vapors or dusts may get into the air, appropriate precautions such as the use of explosion-proof wiring, provision for adequate ventilation, and the banning of all smoking and use of open flames are necessary. Other hazardous chemicals include organic peroxides, such as benzoyl peroxide. These must be kept away from flammable materials and handled with care. Adequate fire protection equipment is necessary. Specific hazards may have individual protection, automatic in action. For example, tanks may be protected by carbon dioxide, nitrogen, or foam provided automatically through special nozzles if a fire breaks out. Proper types of fire extinguishers properly located are necessary.

Health Hazards

Health hazards result mainly from the use of acids, solvents, and plasticizers which may cause injury to the person from contact or inhalation. Here again keeping volatile materials in tight containers and providing for adequate ventilation are desirable. Protective devices such as goggles and gas masks may be necessary.

In General

Safety is achieved as in any other industry by proper design and layout of the manufacturing plant, together with proper operation. Proper operation includes adequately trained personnel, adequate understanding of the potential hazards of the materials being processed and the operations involved, good housekeeping, regular inspections, adequate maintenance, and planned safety.

CHAPTER 3

Production of Finished Plastics Products

Most of us think of plastics in the molded form. The many products familiar to the public are made in different types of equipment by techniques adapted to both the equipment and the molding material. In addition, plastic materials have other uses such as in coatings, films, sheets, adhesives, and fibers.

CASTING

The Process

Basically, casting consists of pouring the liquid plastic into a mold and allowing it to harden. Certain variations of casting used for specific applications are generally recognized. *Embedding* is the enclosing of materials, such as biological specimens, in a block of transparent plastic. *Potting* is the impregnating and covering of electrical parts of assemblies for purposes of insulation and protection. *Encapsulating* is similar to potting except that the electrical parts are embedded in cellular or foamlike plastic which completely fills the space within the housing.

Both thermosetting and thermoplastic resins may be cast with all or part of the polymerization occurring in the mold.

38

Certain thermoplastic resins are simply poured into the molds in melted form and allowed to harden. Polymerization in mold may occur at room temperatures or be carried out in ovens or autoclaves. In some cases cooling is necessary to prevent overheating.

Molds

Draw molds are made by dipping a steel arbor or mandrel of the proper size and shape into molten lead. Upon removal the lead solidifies to form an open lead mold. The mandrel is tapered slightly to facilitate removal of it and the castings. *Split molds* are used for shapes, such as those with undercuts, which cannot be removed from a draw mold. The parts are clamped together. *Cored molds* use a metal core to form the inner shape of the piece. Flexible molds are made by building up approximately a 1/8-inch layer of rubber, plastisol, or similar flexible plastic on the surface of a metal, hardwood, or porcelain model. *Plaster molds* made by casting plaster of paris around a model are also used. In *potting* the mold may be a casing which becomes an integral part of the assembly. Removable molds are made of various materials such as plastics, metals, and glass. In *encapsulating* the mold is normally the casing.

Applications

Standard shapes, such as plates, rods, and tubes, are frequently made by casting. Casting is also used where the small number of pieces of a desired plastic item does not justify the expense of a mold for injection or compression molding. Large pieces can sometimes be better made by casting than by molding.

COMPRESSION MOLDING

In compression molding the molding material is placed in the cavity of a hot mold and pressure applied through the upper or plunger part of the mold to cause the softened plastic to conform to the contour of the mold.

Compression molding is well suited for thermosetting plastics where the final polymerization occurs in the mold itself. Molding under pressure assures good contact between the resin and any filler and reduces the disruptive effect of evolved water vapor or gases, thus producing a denser and stronger finished product.

Since the plastic sets up thermally in the mold, the finished piece can be discharged without cooling.

Thermoplastic materials can also be molded in compression molds. Since they do not harden with heat it is necessary to cool the mold before removing the molded piece, thus slowing down the operation.

Molding Presses

Since molding pressures are high, as much as 800 pounds per square inch of surface, the pressures required by large pieces may be very great. These pressures are applied by direct hydraulic pressure or a combination of hydraulic, air, or steam pressure with a toggle mechanism. The mold cavity is commonly mounted on the lower movable platen, and the plunger, or force, attached to the top fixed platen.

Molds

Three general types of molds are usually recognized: *positive, semipositive,* and *flash* (see Figure 3.1). In the *positive* mold the limitation on the downward travel of the plunger or force is the material in the mold. Thus the thickness of the molded part depends upon the amount of material in the mold. In the *flash* mold the comparatively loose fit of the plunger allows the excess material to squeeze out the sides as flash. The *semipositive* mold has the travel of the plunger limited by the "land" or stop. If there is an excess of molding material in the mold some of it may be pressed out as flash. The flash mold is the simplest and

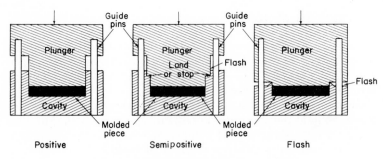

Fig. 3.1—Types of compression molds.

cheapest type and is suitable for flat pieces. The other types are better where greater depth or "draw" is necessary. Molds for complicated pieces may have cavities made in two or more parts and have to be assembled and disassembled by hand. Molds may be single or multiple cavity, producing one or more pieces at a time. Most compression molds are machined from special steel and then hardened. Frequently they are chromium plated. Special alloys may also be used. Multicavity molds are frequently made by pressing a hardened steel hob having the shape of the finished plastic piece into a block of mold steel, forming the cavities. The mold may be heated from the platen, or directly, usually by steam. Electrical resistance and dielectric heating are also sometimes used.

Preparation of the Molding Material

If the molding "powder" is moist it may need to be dried to prevent the formation of surface defects or porosity from the resulting steam. The powder may be preformed, that is, pressed into pellets or disks of convenient size to reduce bulk density and facilitate charging the mold. One or more preforms may be used alone or with added powder.

Preheating the molding powder and preforms almost to the polymerizing temperature may be desirable, especially in molding large pieces. Unless this is done, the outside of the piece may be thermally set before the center has softened sufficiently to flow properly. Preheating may be done in shallow trays in steam or electrically heated ovens or by dielectric heating.

Molding Operations

In *manually operated* presses the molds are filled and the finished product removed by hand. The molding time is controlled by hand-operated valves. The mold parts are commonly attached to the platens and not removed. Molds used for complicated pieces may be removed from the press and disassembled to remove the pieces. The common *semiautomatic* press controls the time cycle automatically. Finished pieces are usually ejected from the mold by knock-out pins, but must be removed by the operator. The operator fills the cavities and starts the press. The *fully automatic* press controls the time, temperature, and pressure automatically and needs no attention except filling the hopper with molding powder (see Figures 3.2, 3.3, 3.4, and 3.5).

Fig. 3.2—An automatic press. (Courtesy F. J. Stokes Corporation)

INJECTION-TYPE MOLDING

Injection, jet, and *transfer molding* have one operation in common: the injection or transfer of molten plastic material from a melting chamber into a mold where it hardens. In true injection molding the thermoplastic material hardens upon cooling. In jet and transfer molding the thermosetting material hardens thermally in a hot mold.

Fig. 3.3—Automatic press starts closing. (Courtesy F. J. Stokes Corporation)

Injection Molding

The thermoplastic molding powder is fed from a hopper into a cylinder where it is softened by heat, usually electrical. Formerly the softened plastic was pushed by a hydraulically operated ram or piston into the water-cooled mold where it solidified. To improve on what would have been rather poor heat transfer, a *torpedo* or spreading device was placed between the

Fig. 3.4—Open press with molded parts raised by ejector pins and with comb under them. (Courtesy F. J. Stokes Corporation)

end of the plunger travel and the outlet nozzle. Other equipment, including auxiliary preheating or preplasticizing cylinders, has been developed, but the reciprocating screw machine is rapidly being adopted. In this machine (see Figure 3.6) the ram is replaced by a rotating screw. The mixing and shearing action of the screw produces a uniform product which is heated to a considerable extent by the mechanical work input. Heat trans-

Fig. 3.5—Parts being discharged by comb. (Courtesy F. J. Stokes Corporation)

fer from the heating elements is improved. The melted material is moved to a space at the end of the screw next to the nozzle. The mold is filled by moving the screw ahead like a ram. After filling the mold, the screw retracts and begins heating and mixing another shot of plastic.

The molds are commonly of steel and made in two parts which are held together by either mechanical or hydraulic pres-

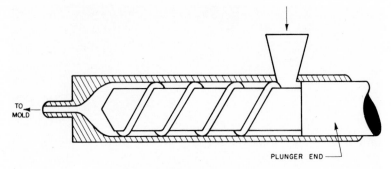

Fig. 3.6—Injection mold feeding device.

sure during filling and solidification of the plastic. While machine overhead tends to be high, this is offset in quantity production by short cycles and low labor costs.

Jet Molding

Jet, or *injection transfer,* molding is a variation of injection molding using thermosetting resins. The molding material may be heated and plasticized in an auxiliary unit using a horizontal extruder type screw to place a metered charge in the transfer pot from which it is injected into the closed mold by a plunger. Another method is to heat and plasticize in an injection chamber from which it goes directly to the mold.

Regardless of the equipment used, it is necessary to control the time and temperature so that the final cure does not occur outside the mold, but occurs in the minimum time once the molten plastic is in the mold. These methods are automatic and operate at high capacities, producing good products at low costs.

Transfer Molding

Transfer molding is a variation of compression molding designed for molding articles from thermosetting plastics with inserts. In compression molding when pressure is applied on the molding powder before it has become highly fluid, any inserts in the mold cavity may be moved out of position, resulting in a faulty product. In transfer molding enough molding powder for one mold is melted at a temperature just below that resulting in polymerization and is forced, or transferred, from the pot into the mold cavity by a plunger.

EXTRUSION

Extrusion is a process of forcing a plastic material through a die to continuously produce such products as rods, tubes, films, bars, filaments, and a variety of other elongated shapes with uniform cross sections. More thermoplastic material is processed by extrusion than by all other forming methods. *Dry hot* extrusion is the type ordinarily referred to simply as "extrusion." Extrusion of plastic solutions will be referred to as "wet cold" extrusion.

Dry Hot Extrusion of Thermoplasts

In this process the thermoplastic molding material is fed from a hopper into a hot cylinder (see Figure 3.7). The heat softens the material and it is forced by one or more spiral screws through the cylinder and out through a die orifice. The die forms the cross-sectional shape of the continuous mass of plastic as it passes through. The extruded product is carried by a conveying device through either the air or a water-cooling bath to harden

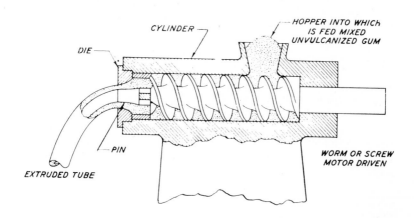

EXTRUSION

Fig. 3.7—Extruder. (Courtesy Phillips Petroleum Company)

it. The cylinders may be heated by electricity, hot oil, or steam, and cooled by water.

Extruded products vary in size from fine nylon textile fibers to plastic pipes 18 or more inches in diameter. Various rods, tubes, bars, and similar cross sections are produced as standard shapes for later fabrication into finished products. Inner tubes for tires are produced by extrusion. Wire is covered with a plastic coating by attaching a crosshead with a die which extrudes at right angles to the cylinder. The wire enters the crosshead at the rear of the die, passing through it in a straight line.

Two methods are used for extruding film: blown tubular extrusion and flat extrusion. In the blown method a large bubble-like tube is extruded using warm air to expand the tube. This tubing is excellent for bag production. If a flat sheet is desired, the tube may be split. In flat extrusion the plastic is extruded through a die with a long and very narrow opening. Blown film, because of both the transverse and longitudinal stretching, tends to have more uniform properties than the flat product which is stretched only lengthwise. In a recent development the flat extruded film is stretched transversely after extrusion as it passes through a heated chamber.

Extrusion has been combined recently with other methods of plastic forming. Extruded sheets or films are continously *vacuum-formed* while the sheet is still warm. Extrusion has been combined with blow-molding for high speed production of plastic bottles. In *extrusion injection* molding the hot plastic from the extruder is pumped directly into molds. Film from an extruder may also be fed directly to rolls for calendering operations.

Extrusion of Thermosets

Extrusion of thermosetting plastics is a recent development. The molding powder is forced into a long, tapered die by repeated strokes of a ram. The die is heated in zones. The charge becomes heated and softened as it moves through the die under high pressure, becoming cured as it reaches the outlet end. A wide variety and size range of profiles are being extruded.

Wet Cold Extrusion

Plastic solutions are extruded to form filaments for textile use and films. Solidification of the plastic filament or film results from evaporation of a solvent, as acetone from cellulose acetate, or by chemical reaction in a hardening bath, as xanthate in an acid bath to form cellophane or rayon.

Filaments are formed by forcing the solution through a spinnaret, a metal plate with one or more holes, to produce either a monofilament or a multifilament. Film is formed by extruding through a narrow slit in a die. The film process is sometimes referred to as "casting," probably a carry-over from the method of forming films from a solution by running the solution onto the surface of a large wheel where the solvent evaporates.

BLOW-MOLDING

Sheet Blowing

One of the earliest methods of forming hollow plastic objects was by blow-molding of thermoplastic sheets. A metal mold was split so that two softened sheets could be clamped between the parts. Air or steam was then blown in between the sheets, forcing them out against the inside of the mold. The sheets were then fused or cemented together around the edges.

Extrusion Blowing

This is a later method in which extruded tubing is expanded within a mold (see Figure 3.8). In the continuous method, hot tubing from an extruder enters an open mold attached to a revolving turntable. The mold closes, air is turned on, expanding the tubing against the mold, and the mold opens, ejecting the finished bottle. Another method uses a blank or preform produced by injection molding and expanded in a mold by injected air pressure. A recent development is a machine which blows the bottle and fills it and caps it, all in one integrated sequence.

PRODUCTION OF SHEETING AND FILMS

Practically all thermoplastics can be produced in the form of films and sheeting. "Films" are less than 10 mils and "sheeting" over 10 mils thick. Both may be produced by extrusion. Sheets of acrylic polymers are made by polymerizing the monomer between glass plates.

PLASTIC ——————

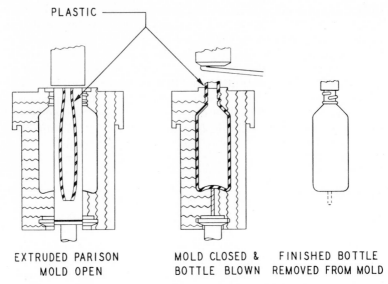

EXTRUDED PARISON   MOLD CLOSED &   FINISHED BOTTLE
    MOLD OPEN   BOTTLE BLOWN   REMOVED FROM MOLD

Fig. 3.8—Blow-molding. (Courtesy Phillips Petroleum Company)

Calendering

The plastic material, with any added plasticizer and color, is passed between a series of heated rollers, finally emerging from the last pair as film or sheeting. The term is also applied to the application of a solid but soft plastic film to one or both sides of a sheet of paper or cloth between heated rolls.

Casting

Sheet casting is a different process from the casting process used as a molding procedure. A solution of the plastic is flowed onto a polished, slow-moving drum or wheel or onto a stainless steel belt where the solvent evaporates, leaving the film or sheet of solid plastic.

Sheeting From Blocks

Sheets of cellulose nitrate are sliced in desired thicknesses from 5 mils upward from solid blocks of the plasticized material.

Laminating

Films are laminated with other films to produce a product with better mechanical properties, better chemical resistance, or better appearance. Plastic films are not only laminated with other plastic films but also with paper, cloth, and metal films such as aluminum. Adhesives are usually applied to the film from rolls, with the amount of adhesive controlled by a pair of rolls or a doctor blade. The adhesive may be dissolved in water or an organic solvent or in an emulsion. The adhesive layer remaining after evaporation of the solvent is usually about 1 mil thick.

Another method, extrusion lamination, consists in extruding a film directly onto a film of plastic or a web of cloth or paper as the latter passes down between two rolls. A large cooling roll may be included together with draw rolls. No adhesive is necessary. If two or more webs of cloth are to be laminated and coated with a resin, they can be carried down through the molten resin or a resin solution and brought out to be pressed together between two rolls. If desired, a sheet of plastic may be laminated to the surface.

SHEET THERMOFORMING

Sheets of thermoplastic materials soften when heated and harden when cold. While hot, some of the thermoplastic sheets can be reshaped by air pressure or other mechanical means, and hold the new shape when cold. Such an operation is known as thermoforming.

Vacuum Forming

In vacuum forming, a vacuum between a hot thermoplastic sheet and a mold causes the air pressure to force the softened sheet against the mold (see Figure 3.9). The mold may be either convex ("male") or concave ("female"). The sheet of plastic material may be heated in a gas or electrically heated oven to proper softening temperature and then transferred to the vacuum machine. An alternate method is to heat the sheet with infrared lights while it is clamped in place on the machine.

The vacuum machine consists essentially of a mold box containing the mold and with vacuum connections at the bottom.

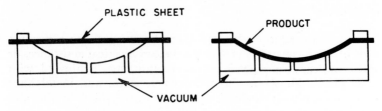

PLASTIC SHEET

PRODUCT

VACUUM

Fig. 3.9—Vacuum forming.

Clamps are provided to hold the sheet tightly against the top of the mold box. More complicated continuous machines can be connected directly to an extruder to utilize the hot output directly.

Drape Forming

Drape forming is a variety of vacuum forming in which the clamped sheet is first pulled down with the outer edges on a level with the lowest part of the mold. Vacuum is then applied to complete the operation. The male mold may have female, or depressed, sections.

Pressure Forming

In some sheet-forming operations it is desirable to have more air pressure than provided by a vacuum. This is done by applying direct air pressures up to 150 psi. A mechanical "helper" is sometimes combined with air pressure in female molds.

Mechanical Forming

Thermoplastic sheets heated to proper temperatures can be formed on sheet-metal equipment. Sheets can be pressed between matched dies similar to compression molds. Cuplike shapes can be formed between a plug and a ring.

COATING METHODS

Liquid Coatings

Paints, enamels, lacquers, and other coating products in which plastic materials form one or more constituents are applied by the familiar methods of hand *brushing,* hand *rolling,* and *spraying.* In factories objects are frequently coated by *dipping.* This is often done by conveying the pieces mechanically down into and up out of a bath of the coating material. Continuous sheets of cloth and paper are also coated in this manner. Instead of passing the sheet down into the bath the coating material may be carried up to it by a roll revolving under it and partially submerged in the bath.

Slush coating is used to cover the interior of an object which cannot be reached by brush or spray. The coating material is poured into the cavity and the object rotated to spread it uniformly over the surface. Any excess is allowed to drain out.

Molten Plastic Application

Two methods of molten plastic application are used: *spraying* and *fluidized-bed* application. In *spray* coating, powdered resin is blown onto the surface to be coated by a special burner nozzle where it is heated by either a gas flame or electrical resistance wires. The melted resin particles coalesce on the surface, forming a continuous coating.

In the *fluidized-bed* method, a fluidized bed of powdered plastic is maintained by a current of air or inert gas at a temperature just enough below the melting point to prevent the particles from adhering together. Hot metal objects to be coated are lowered into the plastic bed and then raised out of it. Upon coming into contact with the hot object the plastic softens sufficiently to adhere to it and the residual heat in the object causes the plastic to flow over the surface after removal from the fluidized plastic bed.

Metal Plating Plastics

Thin metallic coatings are sometimes deposited on plastic surfaces for decorative purposes. One of the early methods was to deposit the metal film from an aqueous solution of its salt by chemical reduction. *Vacuum metallizing* is commonly used. The clean plastic part is frequently coated with an organic base coating of paint or lacquer to form a smooth surface. The pieces are placed in a vacuum chamber where a metal, usually aluminum, is evaporated almost instantly by electrical resistance heat developed in the metal. The aluminum vapor coats the plastic, forming smooth, uniform surfaces. After removal from the vacuum chamber the coating is given a protective covering of lacquer, either clear or some desired color. *Electroplating* is used to coat ABS plastics parts for replacing zinc die castings and other metal parts. The plastic surface is usually etched first with a mixture of chromic and sulfuric acids to slightly roughen the surface. The surface may then be "sensitized" with tin salts or "activated" in a noble-metal salt solution, followed by plating with copper or nickel in an electroless plating bath. Chromium may be plated over the copper or nickel to the desired thickness. Most of the thermoplastics have been plated experimentally if not commercially.

Hand Lay-up

The extensive production of glass-reinforced plastic products has resulted in the development of several molding methods used mainly, but not necessarily exclusively, in this industry.

In *hand lay-up* the reinforcing material, usually in the form of glass fiber mat, is placed by hand on the surface on a mold and impregnated with liquid resin by spraying or hand brushing (see Figure 3.10). The surface is smoothed by hand and the resin

Fig. 3.10—Hand lay-up of a glass-reinforced polyester boat. (Courtesy Shell Lake Boat Company and Reichhold Chemicals, Inc.)

allowed to set up at atmospheric pressure, in an autoclave under air pressure, or with bag pressure (see Figure 3.11). Heat may be supplied from the sun, or artificial heat such as infrared lights may be used. Molds may be male or female and made of wood, plaster, metal, or plastic. A parting agent is used on the mold surface to prevent the plastic from sticking.

Equipment costs for this method are low but labor costs are high, making it useful for limited production. It has been used in the production of small boats, sports car bodies, and similar items using polyester resins and glass fiber.

Preforms and Mixes

In the more complex moldings the use of mats as in hand lay-up is not so satisfactory since it may result in tearing, wrinkling, and unsatisfactory glass distribution. To avoid these difficulties three types of preforms have been developed: *plenum chamber, direct fiber,* and *water slurry.*

In the *plenum chamber* method, rovings, after being cut, fall into a plenum chamber onto a perforated screen. Air is exhausted from under the screen. Powdered resin may be applied by a vibrator feed, or liquid resin may be sprayed on. In the *direct fiber* method the cut roving is blown out of a hose onto a screen. At the same time liquid resin is sprayed onto the fiber glass. In the *water slurry* method the chopped fibers are formed

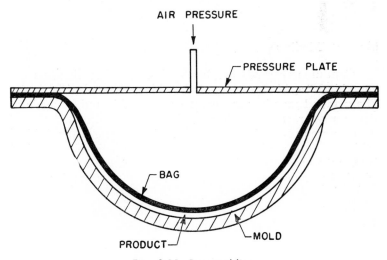

AIR PRESSURE

PRESSURE PLATE

BAG

PRODUCT

MOLD

Fig. 3.11—Bag molding.

into a slurry with about 20 per cent cellulose fibers in water. The slurry is then sucked down over a perforated screen mold.

Instead of preforms, *"pre-mixes,"* mixtures of chopped fiber and resin corresponding to molding powders, may be used. Mats or fabrics may be impregnated before molding with resin to produce *"pre-preg"* materials. *"Pre-compounded"* materials may contain more than one resin plus fillers other than glass. Colorants can also be added.

Precision Molding Methods

In the flexible plunger method the piece is formed in a presion-made heated metal cavity by a flexible plunger, usually of rubber, designed so that it contacts the lower part of the mold first. As the contact area expands, the resin and the air are forced upward through the reinforcing material. Pressures usually range from 50 to 150 psi. This method allows for some irregularities in the thickness of the mat or preform and produces a piece with a smooth outside surface and no voids.

In the *vacuum injection molding* process resin is forced up by air pressure from a trough into a dry lay-up of reinforcing material held in a mold. This technique has been used to produce boats and tanks. *Matched die molding* is a variation of compression molding in which mats or preforms are molded in heated dies. Pressures used run from 100 to 200 psi. Moldings are uniform and strong. This method is excellent for high volume operations.

Filament Winding

Continuous strands of glass roving or other filaments are wound by a machine on a mandrel and impregnated with a resin. The mandrel may be permanent or removable and be made of plastic, plaster, steel, or other metals and may be rotated around one or more axes during the winding. The shape of the finished product may be a cylinder, sphere, or conical body of revolution. Among the products made by this technique have been radomes, spherical pressure vessels for gases on aircraft, and rocket motor cases.

Centrifugal Molding

A fiber-glass mat formed around a pipe is inserted into a hollow mandrel which is revolved while resin is sprayed over the inner surface from a central pipe. The centrifugal force spreads the resin uniformly over the glass fiber and forces out any air. These moldings have been used as water-softener tanks, torpedo-launcher tubes, and external-pressure tubes.

Centrifugal molding is also applied to thermoplastics such as polyethylene. The molten plastic is formed into a layer on the inside of a rotating metal cylinder, forming large-sized pipe.

Sprayup

Sprayup is a method of simultaneously spraying catalyzed resin and continuous or chopped-glass roving onto a mold surface. The resin and catalyst may be projected in separate streams or mixed in the head of the spray gun just prior to ejection to avoid premature gelling. After the spraying, the glass resin mixture is rolled to compact it and eliminate air bubbles.

EXPANDED PRODUCTS

Types

Expanded plastic products having a cellular structure can be made from rigid plastics, flexible plastics, and elastomers. Two general types are manufactured: *open* (interconnecting) or *closed cell* structure.

Production

Five general methods of producing a cellular structure are used: (1) mixing with the liquid or plastic mass a *chemical blowing agent,* such as sodium bicarbonate, which produces a gas by a chemical reaction; (2) mixing in a cross-linking agent which produces gas; (3) adding a volatile solvent which forms gas when heated; (4) mechanically mixing in air or other gas; and (5) adding tiny expanded spheres to a suitable binder material, producing "syntactic" foam.

Forming

The expanded product may be formed into final shape in either an open or closed mold, with or without heat. It may be formed *in situ,* surrounding members of the final structure forming the mold. It may be extruded onto a moving belt to form slabs or sheets. The foamed material may be sprayed onto a surface. Expanded beads, such as those of polystyrene, are molded in a steam-heated mold. A perforated mold may be placed in a steam chest, and steam injected directly into it. Another method is to place the mold in an autoclave under steam pressure.

LAMINATING

Laminates consist of two or more sheets or films, one or more of plastic or plastic-coated or impregnated filler, bonded together. The filler may be paper, cloth, sheet asbestos, or wood. Sheets or films of metal may also be incorporated. It has been customary to speak of laminates molded at less than 100 psi as *low-pressure* laminates as opposed to *high-pressure* laminates molded at higher pressures. Glass-reinforced products and laminated films may be considered as low-pressure laminates. They are discussed elsewhere.

High-Pressure Laminates

Sheets of resin-impregnated filler are stacked in piles between heated platens of a hydraulic press and pressed into sheets. Tubes are made by forming over mandrels and curing. Plywood is a laminate of thin layers of wood veneer bonded together by a resin.

THERMOFUSION

In the *thermofusion* process a powdered thermoplastic is heated in a sheet-metal mold until it fuses. The heating is carefully controlled in an oven. Cylindrical tanks up to 450 gallons capacity and rectangular tanks up to 500 gallons capacity have been produced from polyethylene. Wall thicknesses are limited to about ⅜ inch. The products are said to be practically stress-free and not subject to cracking.

SHELL MOLDING

The term *shell molding* has been applied to two processes. The first of these, also called "slush molding," is described elsewhere. In foundries shell molds are used for accurate casting of metal parts. A thermosetting resin, usually phenolic, is mixed with fine sand and applied in a thin layer over the pattern, followed by heating to fuse and thermally cross-link the resin. The "shell" is then stripped from the pattern and is ready for use.

FABRICATION TECHNIQUES

Not all plastic articles can be molded into final form. Many can be built from sheets or other standard forms. Three general methods are used for fastening these together: *cementing, welding,* and fastening *mechanically.* The latter two methods resemble those used in joining metals.

Cementing Thermoplastics

Three types of cements are used for bonding thermoplastic materials: *solvent, dope,* and *polymerizable cements. Solvent* cements are simply solvents or solvent mixtures which dissolve or soften the surfaces to be joined. Thus a cement is produced in which a portion of the surfaces to be bonded becomes the cementing medium. Low-boiling solvents set readily but may cause undesirable surface crazing. *Dope* cements are simply a solution of the plastic to be bonded in a solvent. They are useful for joining imperfectly fitted parts. *Polymerizable cements* are reactive monomers compounded with necessary catalysts and promoters so that they will polymerize below the softening point of the materials to be joined. The monomer must be compatible but not necessarily identical with the polymer being cemented.

Welding

Most thermoplastics can be heat welded by several techniques. In general, plastics most highly polymerized produce the strongest joints. Welding temperatures must be high enough for proper softening but low enough to prevent thermal degradation or burning. *Hot gas welding* of plastics resembles gas welding of metals except that hot gas, not a flame, is used. Air or an inert gas is fed through a torch where it is heated to the proper temperature by either gas or electricity. The welding rod is usually of the same composition as the material being joined.

In *induction* or *electronic welding* heat is induced by the action of a high frequency electric field on either a metal insert or a sealing die. Welding, using a metal insert at the interface of the surfaces to be joined, is rapid. The welding occurs in a small area around the insert so that the strength of the weld may be slight. For welding or *heat-sealing* films, induction-heated metal dies between which the films are clamped may be used efficiently. In *spin* or *friction welding* one of the two pieces to be joined is rotated against the other, producing enough heat to melt the surfaces. When sufficient heat has been developed, the rotation is stopped and the pieces are pressed together until cooled. In *heated-tool welding* the parts to be joined are heated by an electrically heated resistance strip, a hot plate, or a modified soldering iron, and the softened areas brought together and cooled under pressure.

Mechanical Fastening

Plastic parts may be joined mechanically with such standard fasteners as machine screws, bolts, and rivets. Machine screws of either metal or plastic may be screwed into threads in the plastic piece. These threads may be molded, drilled, and tapped, or may be in inserts. Metal self-tapping screws are also used. Metal and plastic rivets of conventional and special design are used. Metal spring clips and nuts are designed for rapid fastening. Other joining techniques similar to those used with metals are swaging and press or shrink fitting.

CHAPTER 4

Natural Plastic Materials

Certain natural plastic materials were known and used long before synthetic plastic resins were developed. While at the present time these are relatively not as important as the synthetic products, they are still widely used in industry. About the only processing treatments given these materials originally were mechanical, largely to remove gross impurities. It seems desirable to consider both the natural materials and their modifications and derivatives together. Cellulose and rubber, which logically fall in this general group, will be considered separately.

PLANT RESINS

These resins originate as solutions of complex aromatic acids and complex saturated organic compounds called *resenes* in essential oils, such as terpenes present in plants. The solutions originally exuded from the trees as the result of accidental breakage of twigs or limbs or as the result of insect attack. On the exposure to the air they gradually hardened to form the solid resin, partly by evaporation of the essential oil and partly by oxidation and polymerization. The physical and chemical properties of resins frequently changed slowly so that the freshly formed product might vary considerably from that which had exuded months or years before. Some of the resins became covered with soil and developed into the so-called *fossil* resins. Some resins are

61

now recovered by tapping the trees so as to obtain a greater yield.

Four major groups of natural resins are commonly recognized: (1) soft, natural resins, such as rosin and dammars; (2) semifossil resins, such as East India and batu; (3) fossil resins, such as copals, kauri, and congo; and (4) a miscellaneous group, including such resins as elemi, mastic, and sandarac.

Rosin

Rosin is a product of several varieties of pine trees and is produced and used in the largest quantity of any natural resin. It is composed of about 90 per cent complex organic acids, largely abietic (see Figure 4.1) and related acids, and about 10 per cent nonacids such as resenes and terpenes. When the pine tree is tapped a mixture of about 68 per cent rosin, 20 per cent turpentine, and 12 per cent water is obtained. The turpentine and water are distilled off, leaving what is known as *gum rosin*.

Wood rosin is extracted from pine stumps remaining after lumbering operations. The stumps are removed by bulldozers, chipped, and the rosin extracted by a light aromatic-aliphatic mixed solvent. The coloring matter is partially removed from the solution by adsorption by a bleaching earth or clay or by

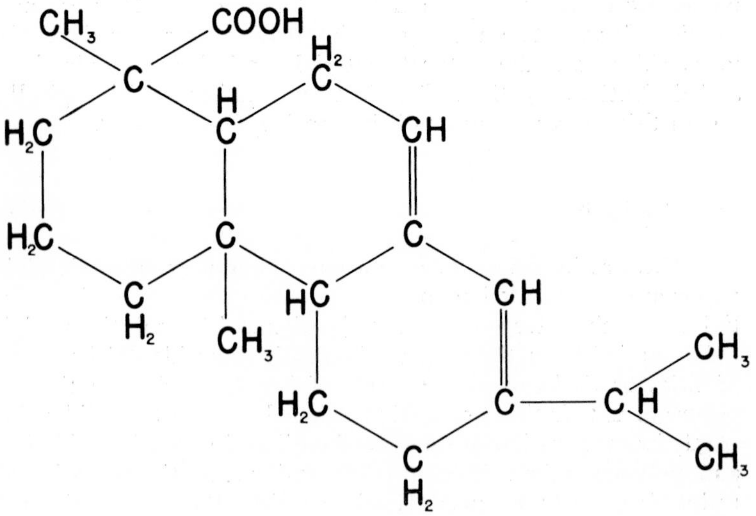

Fig. 4.1—Abietic acid.

liquid-liquid extraction with furfural. *Tall oil rosin* is separated from crude tall oil obtained from kraft cook liquor used in the production of paper from pine wood.

The rosin acids can be esterified by heating with various alcohols. The methyl ester is a heavy, viscous liquid used in mastic compositions, in asphaltic impregnants, as a tackafier in adhesives, as a drying oil extender in linoleum, and as a varnish constituent. The glycerol ester, known as *ester gum,* melts at 197.6° F and has long been used as a lacquer and varnish resin. When pentaerythritol is substituted for glycerol, a resin with a melting point of 239° F is produced. Varnish made with this resin dries more rapidly than that made with ester gum and has greater resistance to water and alkali. Hydrogenated methyl esters are used as plasticizers. The sodium soap of rosin has extensive use as a paper size.

Damars

Damar, or dammar, is secured by tapping a pine tree which grows in Indonesia and on the Malay Peninsula. It varies from colorless to yellow, bleaching upon exposure to ultraviolet light. While it can be used in molding powders, its main use is in paint, enamels, and varnishes. A soft grade of *manila* resin results from tapping *Agathis* trees. It is used as a sizing material and a shellac substitute.

Copal Resins

Copals are fossil resins. Of these the *congo* resin from Africa is one of the hardest natural resins. *Kauri* is a similar resin from New Zealand. Fossil grades of manila resins are also found. These resins are heat cracked to make them soluble and are used in various coatings. *Amber,* a yellow to almost black fossil resin, is also used in jewelery.

Miscellaneous Resins

A miscellaneous group of resins derived from trees is used in varnishes, adhesives, and printing inks. This includes gum accroides, elemi, mastic, and sandarac. *Lignin* (see Figure 4.2) is the encrusting and binding material on most cellulose in plant tissues. It is available in modified forms as a by-product of the

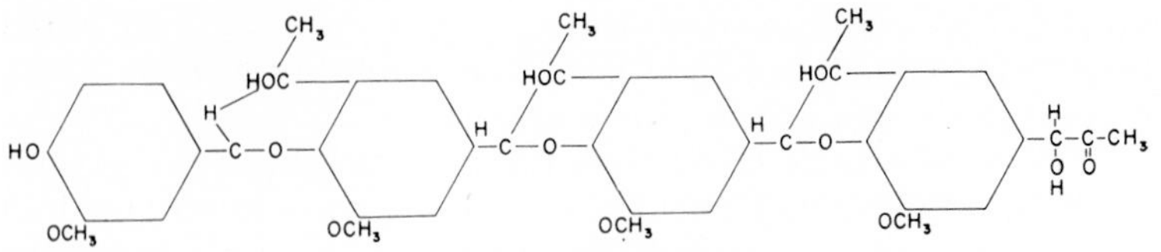

Fig. 4.2—Lignin formula.

paper industry. The lignin molecules are very large, with aromatic rings with methoxy and hydroxyl groups. Lignin appears to be a promising raw material for plastic resins, but has only a limited use mainly as an extender in phenolic resins and rubber.

PROTEINS

Proteins form a large group of complex nitrogenous compounds of both animal and vegetable origin. The basic building blocks of the proteins are the amino acids which have the general formula, $RCH(NH_2)$ COOH, where R may vary from a hydrogen atom in glycine to a long carbon, hydrogen, and oxygen chain or a ring structure in the more complex acids. The nitrogen may also occur as a part of a ring structure. These amino acids are combined into still larger molecules to form the proteins with molecular weights from 30,000 into the millions.

Animal Proteins

Silk and *wool* are natural animal proteins used as fibers without further chemical treatment. Animal hides are chemically processed to produce *leather,* which is not normally considered a plastic product.

Casein can be precipitated from skimmed milk by rennin, an enzyme, or by an acid such as hydrochloric. Plastic products were originally made from casein by mixing with about 35 per cent water, heating and extruding into an approximately 5 per cent solution of formaldehyde. The action of the formaldehyde is probably similar to that in tanning, reacting to block the hydrophilic groups, such as amino and carboxyl, and resulting in some cross-linking. After hardening, the product was machined to the desired shape. The hardening process was very slow, taking from a minimum of a week to even as much as a year, depending upon the thickness of the piece. This limited the use to small objects such as buttons, buckles, and small jewelry items.

Later it was found that the addition of about 2 per cent alum made it possible to machine the extruded product before hardening. Buttons and similar small objects are turned out in automatic machines and then hardened in the formaldehyde bath. Dyes and pigments may be added to the casein and water before extruding to provide the desired color in the product. Casein plastics are also available in sheets up to $\frac{1}{4}$ inch thick, rods up to $\frac{3}{4}$ inch in diameter, as well as tubes and disks. Casein plastics compare favorably with most thermosetting plastics in tensile, compressive, and impact strengths. They can be machined readily. While resistant to many organic solvents they are decomposed by acids and alkalies. Their major limitation to extensive use is their high moisture absorption and dimensional instability.

Textile fibers are made by extruding a dispersion of casein in a sodium hydroxide solution into a hardening bath containing dilute sulfuric acid, formaldehyde, and glucose. Hardening under a slight tension increases the strength. They are ordinarily produced as staple rather than continuous fiber. In warmth, casein fiber resembles wool, with which it is frequently blended. Its wet strength is low and like wool it is attacked by clothes moths, carpet beetles, and mildew. Competition by synthetic fibers has limited the use of casein fibers in this country.

Vegetable Proteins

Protein from *soybeans, peanuts,* and *corn* has been used in both molded plastics and textile fibers. When the oil in soybeans is extracted by hexane, the residual meal on a dry basis contains about 48 per cent protein, less than 1 per cent oil, about 7 per cent ash, 7 per cent fiber, and 37 per cent carbohydrates. The protein is dissolved in a sodium sulfite solution from which it is precipitated by sulfuric acid. The protein from solvent-extracted peanut meal is removed by a dilute sodium hydroxide solution. The corn protein, zein, is present in the gluten separated from the corn in starch manufacture. It is extracted by isopropyl alcohol and the impurities removed from the alcohol solution by liquid-liquid extraction by hexane. The zein is precipitated from the alcohol solution by adding water.

Fibers have been made from the vegetable proteins by methods similar to that used for casein fibers. While it was found possible to dissolve the peanut and corn proteins in a concentrated urea solution, this proved to be too expensive. The usual procedure is to dissolve the protein in a sodium hydroxide solution and extrude into a coagulating bath containing sulfuric acid and formaldehyde or sodium sulfate. The fibers may be given an additional hardening with formaldehyde after stretching to orient the molecules in the fiber. Protein fibers, in general, are soft, with a warm feeling like wool—which they also resemble in their good resistance to dilute acids and poor resistance to alkalies. They have been commonly blended with wool. The fiber from zein is probably the best of the vegetable protein fibers.

Some soybean meal has been used in blends with phenol-formaldehyde resin to form a molding powder. The hardening action of the formaldehyde added before molding produced a product similar to the casein plastics. However, the large amount of carbohydrate was a disadvantage, probably acting only as a filler and increasing the water absorption of the finished product. Attempts to make a plastic from the meal alone by hardening with formaldehyde or furfural have not been successful. Zein has been made into plastic products similar to those from casein. Formaldehyde reacts more slowly with the zein, making it possible to mix the two together before extruding. Zein plastics are said to be more moisture resistant than casein plastics.

SHELLAC—AN ANIMAL RESIN

Shellac, one of the most important natural molding resins or plastics, is secreted by an insect scarcely 1/40 of an inch long which is found by the thousands upon certain trees, such as plum, in India. The lac is harvested by picking the twigs upon which it is deposited, and grinding, winnowing, and screening to separate the resin from the wood. It is then melted and made into sheets. The solidified sheets of lac are broken into flakes which are known as shellac, the common form of lac imported into this country. Bleached, or white, shellac is made by bleaching it in a sodium carbonate solution with hypochlorite. The bleached shellac is precipitated from the solution with dilute sulfuric acid.

Shellac is a thermoplastic resin which has properties that made it for many years an outstanding material for two major uses: phonograph records and high-voltage insulators. It has been replaced as a record material by other plastics such as vinyl polymers. Nontracking types of synthetic resins are replacing it in the insulation field. Shellac has been an ingredient in many products such as dental blanks, grinding wheels, novelties, poker chips, leather dressing, polishes, paper glaze, gasket cement, match heads, and sealing wax. A solution of shellac in alcohol is used as a clear sealer coat under varnish or as a finishing coat on wood.

ASPHALT AND RELATED PLASTICS

These are solid or semisolid dark or black hydrocarbon mixtures related to petroleum. *Asphalt* is a black, solid material found in naturally occurring deposits in Trinidad, Cuba, Venezuela, California, Utah, Texas, and Argentina. It is also produced from certain types of petroleum as the residual product after removing the lubricating and fuel fractions. It is largely soluble in carbon disulfide and has a flowing point of 170° to 190° F. It has been known and used by man since prehistoric times. It was used as a flooring material as early as 3000 B.C. in Sumeria and as a waterproofing material for swimming pools in ancient India. It was used as a street-paving material in ancient Babylon.

Asphalt is used in paving, waterproofing, cements, varnish, paints, roofing materials, cold molded plastics, and floor tile. *Gilsonite* is a pure form of asphalt with melting points from 270° to 375° F. It is found in veins in the ground in Utah and Colorado. It is used in varnishes, paints, floor tile, roofing materials, wire insulation, printing inks, foundry core compounds, and rubber compounds.

Ozokerite is yellow-brown to black or green and waxlike. It is soluble in benzine, benzol, turpentine, carbon disulfide, and ether. It is used in electrical insulation, sealing wax, inks, dolls, and crayons. *Ceresin,* a purified form, is used in face creams, candles, floor polish, wax matches, and carbon paper.

CHAPTER 5

Cellulose Plastics

Cellulose, which forms an essential part of the framework and other tissues of plants, is our most important and abundant organic raw material. It constitutes over 98 per cent of cotton fiber, as much as 50 per cent of most woods, and about 35 per cent of cereal straws. Cellulose for plastics use is derived from cotton or wood.

Cellulose plastics are sometimes referred to as "natural" plastics. Perhaps it would be more logical to think of them as chemically modified natural polymers in contrast to the "synthetic" products polymerized from chemical monomers.

GENERAL CHEMISTRY

The empirical formula for cellulose is $(C_6H_{10}O_5)_n$. The generally accepted structural formula (see Figure 5.1) shows it made up of β-glucose anhydride units, each with three hydroxyl units, and linked by primary valences through oxygen bridges. Each molecule is composed of a chain of approximately 3,000 units. Cellulose fibers are formed of groups of roughly parallel chains held together by bonding varying from weak van der Waals forces through varying degrees of hydrogen bonding. Thus the fibers show both crystalline and amorphous areas.

Cellulose undergoes the following reactions involving the

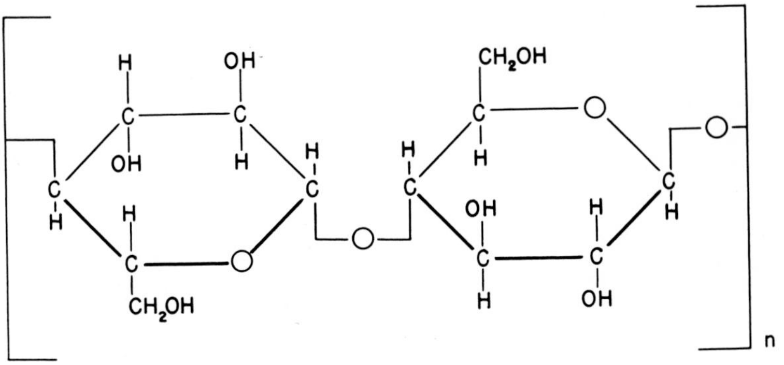

Fig. 5.1—Cellulose.

OH groups: formation of esters such as nitrates, xanthates, acetates, and propionates; formation of alkyl ethers (as ethyl and methyl cellulose), hydroxy ethers, and ether esters; and formation of addition compounds with alkalies, acids, and salts. Cellulose also undergoes hydrolysis to glucose, oxidation, and pyrolysis.

Chemical or "dissolving" cellulose for plastics derivatives varies from an alpha cellulose content of 92 per cent for viscose to 98 per cent for other plastics. Alpha cellulose is the long-chain cellulose characterized by its resistance to 17 per cent sodium hydroxide solution. Cotton linters, too short for textile use, are digested under steam pressure with dilute sodium hydroxide solution followed by chlorine bleaching to produce high alpha cellulose. Paper pulp produced from wood by one of the chemical processes may be converted to high alpha pulp by treatment with sodium hydroxide and chlorine. Sulfite pulp is the most common paper pulp used. The molecular weight of the native cellulose which may be as much as one million may be reduced as low as 150,000 by this processing.

REGENERATED CELLULOSE

Regenerated cellulose is produced by converting cellulose wholly or partially into a derivative and converting it back to cellulose. This results in a product of lower molecular weight and different physical character from the original cellulose. Commercial regenerated cellulose products are cellophane, viscose rayon, and cuprammonium rayon. Vulcanized fiber, although not commonly considered in this category, will, because of its similarity, be included.

Vulcanized Fiber

When high alpha cellulose pulp is treated with a 75 per cent zinc chloride solution, the solution penetrates between the cellulose chains, causing swelling and gelatinization. This swollen pulp is commonly pressed into sheets from 0.005 to 2 inches thick or formed into tubes around mandrels. After thorough washing it is dried. Vulcanized fiber cannot be molded by the usual methods. It can be machined by such operations as cutting, turning, milling, shaping, drilling, and tapping. The sheets can be formed in punch and hydraulic presses by such operations as bending, drawing (see Figure 5.2), swaging, and cupping. Moistening facilitates forming.

Vulcanized fiber has very good physical and dielectric properties but very high water absorption. Standard colors are red, gray, and black. It is used in many electrical applications and in silent gears, wastebaskets, textile bobbin heads, welders' helmets, and gaskets.

Fig. 5.2—Deep draw molding of vulcanized fiber. (Courtesy National Vulcanized Fiber Company)

Viscose Products

Viscose Production. Cellulose reacts with sodium hydroxide to form an addition compound called *alkali cellulose*. This reacts like an alcoholate with carbon disulfide to produce the dithio-carbonic acid (xanthogenic acid) ester. In a simplified form the reaction is:

$$[C_6H_9O_4(ONa)]_n + nCS_2 \longrightarrow [C_6H_9O_4(O-\overset{\displaystyle S}{\overset{\|}{C}}-S-Na)]_n$$

The exact number of xanthate groups introduced varies with the time of the reaction and the amount of carbon disulfide. It is commonly assumed that there is only one xanthate group for each two glucose groups in the commercial xanthate. However, cellulose has been completely xanthated experimentally.

The cellulose xanthate is dissolved in dilute sodium hydroxide to produce the viscose solution. This reacts with sulfuric acid to produce a regenerated cellulose somewhat as follows:

$$[C_6H_9O_4(O-\overset{\displaystyle S}{\overset{\|}{C}}-S-Na)]_n + nH_2SO_4 \rightarrow (C_6H_{10}O_5)_n + nNaHSO_4 + nCS_2$$

The regenerated cellulose has a degree of polymerization of about 600 compared to about 1,200 in purified wood pulp and 3,000 in the native cellulose.

Typical viscose manufacture starts with steeping the sheets of wood pulp having about 90 per cent alpha cellulose content in a 17 per cent sodium hydroxide solution for 40 minutes. Soaking is done in a rectangular tank equipped with a hydraulic ram for squeezing out the solution at the end of the soaking period. The resulting alkali cellulose is next shredded and then goes to storage cans for about 20 hours. From the cans it is dropped through the floor into jacketed octagonal revolving drums known as "barrattes," which hold about 100 pounds. Carbon disulfide is added and the drums revolved to secure good mixing. At the end of the reaction the xanthate is mixed in a blending tank with 5.5 per cent solution of sodium hydroxide. The solution is allowed to age for about 20 hours until it reaches the proper viscosity, as determined by tests. It is then filtered.

Viscose is made into two products: (1) a sheet material known as *cellophane* and (2) a textile fiber known as *viscose rayon*.

Cellophane. In the manufacture of sheet cellophane the xanthate is extruded through a narrow slot into a coagulating solution of 10 per cent sulfuric acid and 30 per cent sodium sulfate. The continuous coagulated sheet is carried on rollers through tanks of water to wash it and through a tank of glycerine to soften it. It is dried on rollers and finally wound into rolls 4 feet wide and about 25 miles long. The dry cellophane is usually treated in a tower with a waterproofing material such as a solution of wax in ethyl acetate or in a solution of cellulose nitrate and is then re-wound. Common thicknesses are 0.001, 0.0015, and 0.002 inches. Cellophane in its original form is transparent and colorless. It is also produced in a variety of colors. Sheet cellophane is used mainly as a packaging material for a wide variety of products. It is also the base for certain self-adhesive types of tape.

Cellophane is extruded into tubes for use as sausage or wiener casings. "Skinless" wieners are stuffed into a cellophane casing on an automatic machine and cooked. The casing is then slit and discarded. Cellulose *sponge* is produced by stirring lumps of sodium sulfate into viscose solution, coagulating the viscose, and washing out the sulfate.

Viscose Rayon. This is the commonest type of rayon and is pro-duced by extruding viscose through fine openings to form fibers which are coagulated in a bath in much the same manner as cellophane. Delustrants such as titanium dioxide or certain oils may be added to the viscose to produce a dull fiber. "High strength" rayon for tire use is made from a higher alpha cellulose, such as cotton linters, and then given a cold stretching to orient the molecules.

Cuprammonium Rayon. A solution is produced by dissolving cellulose in Sweitzers reagent, an ammoniacal solution of copper oxide. A 9 per cent solution is extruded into warm water to produce rayon fibers. Dilute sulfuric acid is used to remove any remaining copper salts which are recovered for reuse.

"Cupra" rayon sells for a higher price than viscose rayon. It is usually produced in very fine filaments for use in high-grade fabrics.

CELLULOSE ESTERS

Various acids react with the OH group of the cellulose to form esters. The commercial products include nitrates, acetates, acetate-butyrates, and propionates.

Cellulose Nitrate

The first cellulose ester plastic produced commercially was the nitrate. Theoretically the reaction is as follows:

$$[C_6H_7O_2(OH)_3]_n + 3nHNO_3 \longrightarrow [C_6H_7O_2(NO_3)_3]_n + 3nH_2O$$

In actual practice complete nitration is not feasible because the product would be too unstable. The dinitrate is the normal plastic material.

Because of its flammable nature cellulose nitrate is made in small batches, 50 pounds or less. High alpha cellulose is stirred slowly with a mixture of approximately 55 per cent sulfuric acid, 25 per cent nitric acid, and 20 per cent water for about one-half hour. As much acid as possible is removed by rapid centrifuging. It is then washed thoroughly with water and finally boiled in water for several hours to degrade to a more stable and more soluble product of lower nitrate content. The boiling, formerly a batch process, is generally carried out continuously in a long tube.

After draining, the remaining water is displaced by ethyl alcohol, and a plasticizer, a stabilizer such as urea, and colorants are added in a dough-type mixer. Camphor, the original plasticizer, is excellent. Its keto groups form strong bonds with the nitrate groups while its bulky molecule spaces the cellulose chains far enough apart to insure flexibility. Sometimes part of the camphor is replaced by orthonitrobiphenyl.

Cellulose Acetate

Cellulose acetate is commonly made by mixing the cellulose with acetic anhydride, acetic acid, and sulfuric acid in a stainless steel Werner-Pfleiderer type of mixer for about three hours. If the triacetate is desired it is precipitated with dilute acetic acid. If secondary acetate is desired the triacetate is partially hydrolyzed by added water before precipitating. The acetate is centrifuged and dried. In another method the reaction between the cellulose and the acetic anhydride is carried out in boiling methylene chloride, using perchloric acid instead of the sulfuric acid as a catalyst.

Cellulose Acetate-Butyrate and Cellulose Propionate

The acetate-butyrate is produced by reacting cellulose with a mixture of acetic and butyric anhydrides. The resulting product has about one acetyl group and two butyral groups per glucose unit of the cellulose. The propionate is produced by reacting propionic anhydride with cellulose by procedures similar to those used for the acetate.

Properties and Uses

The cellulose esters all have good mechanical and dielectric strengths and are available in a wide range of colors. The nitrate (celluloid), because of its heat sensitivity, cannot be injection molded. Compression molding can be carried out only under carefully controlled temperatures. Some extruded products are made. Sheets are commonly made by slicing from solid blocks in thicknesses from 0.003 inch upward. In contrast the other three esters can be readily extruded and injection and compression molded. Unlike the nitrate the acetate is compatible with a wide range of plasticizers such as diethyl and dimethyl phthalates and tricresol and cresyl diphenyl phosphates. The acetate-butyrate is compatible with even more plasticizers than the acetate. Unlike the nitrate the other cellulose esters do not present any great fire hazard. The nitrate has the lowest water absorption, the acetate the greatest. The nitrate deteriorates with age, particularly under exposure to sunlight. The acetate stands exposure to sunlight well but is surpassed in resistance to weathering by both the acetate-butyrate and the propionate. Much can be done in producing acetate formulations for specific applications by control of the molecular weight and acetyl content and the use of proper plasticizers.

The high flammability and limited moldability reduce the utilization of the nitrate. Common uses are piano keys, combs, brush and mirror backs, and novelties. Cellulose acetate has many applications such as electrical parts, including housings for small appliances; ladies shoe heels; knobs; and cutlery handles. Both the acetate-butyrate and the propionate, although somewhat higher priced, can be used for much the same end products as the acetate. In addition, because of better weatherability they can be used in such outdoor applications as automotive parts, fishermen's floats and tackle, and gunstocks. Because of its toughness cellulose acetate-butyrate is well suited for tool handles.

All of the esters are used in film and sheet form. Many small items are vacuum-formed from cellulose acetate and acetate-butyrate sheeting. Films are used in packaging, photographic film, and tapes. Both the acetate and nitrate are used in lacquers. The acetate is extruded from an acetone solution to form textile fibers which have extensive use. Cellulose nitrate can be distinguished from the other cellulose esters and regenerated cellulose by its greater flammability. Unless fire retardant plasticizer has been incorporated, it burns with almost explosive violence. The other cellulose compounds are comparable to paper and wood in flammability. Cellulose acetate-butyrate, if stored away from air, may have a slight butyric acid odor.

CELLULOSE ETHERS

Ethyl Cellulose

Production. Ethyl cellulose is the only cellulose ether used as a plastic material. It is produced by first spraying high alpha cellulose with sodium hydroxide solution to produce alkali cellulose. Either ethyl chloride or ethyl sulfate is then added with some excess sodium hydroxide. The reaction is carried out at around 400° F. Details of time, raw material ratios, and pressures vary with the ethoxy content and viscosity desired.

Properties and Applications. Ethyl cellulose has the lowest specific gravity of the cellulose derivatives. It is tough, has excellent electrical properties, excellent dimensional stability, and low water absorption. It is readily injection molded and extruded. Sheets of it are vacuum-formed.

Ethyl cellulose is used in such applications as football helmets, flashlight cases, furniture trim, refrigerator breaker strips, luggage, and tubing.

Trademark names: Bexoid, Ethocel, Forticel, Plastacele, Sicalit

CHAPTER 6

Phenolics and Aminoplasts

✳

PHENOL-ALDEHYDE PLASTICS

The reaction of phenols and aldehydes to form resinous products, known since its discovery by Baeyer in 1872, was of little interest until the work of Baekeland in 1909. Baekeland developed two techniques that made the successful compression molding of phenolic resin practical: the use of fillers and the use of heat and pressure to prevent the liberated water vapor from rendering the plastic porous.

The reaction between phenols and aldehydes is basically a condensation process in which water is formed as a by-product. If there are only two active hydrogens in the phenol, a chain polymer is formed which is thermoplastic. If more than two hydrogens are available, cross-linking may occur, producing a thermosetting compound.

In phenol (C_6H_5OH) the OH group orients any reacting group either ortho or para, thus limiting the active hydrogens to three (see Figure 6.1). Reaction of an aldehyde with these three can produce a cross-linked structure. In the cresols one hydrogen is replaced by a methyl group. In both ortho cresol and para cresol only two active hydrogens are available, making only linear compounds possible. Meta cresol, on the other hand, has three active hydrogens and can be cross-linked. Resorcinol and 3,5 xylenol can also be cross-linked.

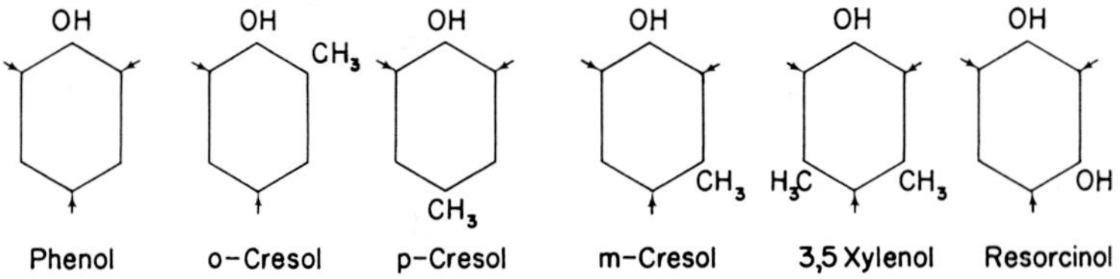

Fig. 6.1—Formulas of various phenolic compounds. Arrows indicate active hydrogens.

Manufacture of Phenolics

One-Stage Resins. In this process an alkali, such as sodium or ammonium hydroxide, is the catalyst. All the formaldehyde for complete reaction with the phenol, about 1.1 to 1.5 moles to one mole of phenol, is added. The constituents are reacted in a steel or nickel-clad steam-jacketed kettle equipped with a reflux condenser and an agitator. The temperature is carefully controlled by water in the jacket and by refluxing the water vapor from the reaction to produce only chain compounds which are thermoplastic.

The reaction is stopped and the water distilled off under vacuum. The resin is mixed on rolls or in a Banbury mixer with filler, color, and lubricant; cooled; and ground to convenient molding "powder" size. When this resin is molded, the reaction between the phenol and the aldehyde is completed in the mold, producing an infusible, cross-linked material.

In a new process under development, a mixture of phenol and formaldehyde is pumped through a series of four horizontal tubular reactors. The sodium hydroxide is added in increasing amounts to each of the reactors. Residence time in the reactors is about 16 minutes. From the reactors the product goes to the cooler. Operating costs are said to be low and product quality uniformly high.

Two-Stage Resins. In this process which uses sulfuric acid as a catalyst, only part of the necessary formaldehyde, about 0.8 mole per mole of phenol, is added at first. The reaction proceeds to completion, producing a "novalac," or thermoplastic resin.

To produce a molding powder the ground resin is mixed with filler, colorant, lubricant, and hexamethylenetetramine. In the mold the "hexa" breaks down into ammonia and formaldehyde, and the resin cross-links to form a thermally set product.

Cast Resins. A higher ratio of formaldehyde, usually 1.5 to 2.5 moles per mole of phenol, and an alkaline catalyst are used for producing cast phenolics. The kettles used for the reaction are usually nickel to avoid any iron contamination. The reaction is carefully controlled to produce a water soluble thermoplastic resin which is acidified with lactic acid or phthalic anhydride or a mixture of the two. Upon acidification the dark resin becomes a light straw color. The water content is evaporated to between 5 and 10 per cent under vacuum. The hot resin is poured out into open molds, usually lead shells. The polymerization occurs in open molds in from 3 to 7 days at 150° to 175° F.

Modified Phenolics. Alkylation of a phenol with such agents as isobutylene, cyclopentadiene, drying oils, and many terpenes in the presence of Friedel-Crafts type of catalyst prior to the reaction with the aldehyde tends to slow the cure of phenolic resins and make them softer and more compatible with drying oils and rubbers. Resorcinol-formaldehyde resins cure at much lower temperatures than the phenol-formaldehyde resins.

Applications of the Phenolics

Molding Powders. Phenolic molding powders are available with a considerable range of molding and finished-product characteristics. In part these differences result from variations in the resins themselves, resulting from different ratios of the primary constituents, or from inclusion of modifying constituents. Greater than these in effect on strength properties are the various fillers which are added in amounts up to 60 per cent or even in extreme cases up to 90 per cent of the molding powder. *Wood flour* is used in general purpose phenolics. The powder is easily molded and the products have good appearance, fair strength, low water absorption, and are low in cost.

Cotton flock, sisal fibers, and *shredded rags* produce molding powders with a high bulk factor and more difficult to mold than those with wood flour. The products have better strength characteristics, particularly *impact* strength. *Asbestos* filler produces a denser product with a lower water absorption and better heat resistance than general purpose products. *Mica* imparts good electrical characteristics and heat resistance. *Graphite* adds resistance to some chemicals and gives the product lubricating properties. *Glass fiber* gives excellent strength.

Pressure Molded Products. Phenolic resins with fillers are molded readily by compression, transfer, and jet molding into a wide variety of products. The molded shapes are thermally set as they come from the mold and cannot be softened by heat. The molded products have good strength, excellent aging properties, fair electrical properties, are nonflammable, and are low in cost. They are available only in dark colors, have a high density, and poor arcing and tracking resistance.

Molded products vary in size from tiny electronic parts to a 285-pound intricately contoured filter plate, 24 square feet in area. Many pieces are molded in multicavity molds on automatic machines; others, because of complexity, by manual operation. The many applications include common electrical parts such as light sockets, receptacles, and switches (see Figure 6.2); instrument cases; telephone bases and handsets; and radio and television cabinets. Heat-resistant knobs and handles for household cooking utensils and appliances are usually molded from phenolic resins. Other products are washing machine and dishwasher impellers and detergent dispensers (see Figure 6.3), caster

Fig. 6.2—Switch parts for an automatic water softener. (Courtesy Durez Plastics Division, Hooker Chemical Corporation)

Fig. 6.3—Filter-dispenser unit for an automatic washing machine.
(Courtesy Durez Plastics Division, Hooker Chemical Corporation)

wheels, camera cases, and vacuum cleaner housings. Phenolic resins are also used in shell molding of metals in foundries. Recently phenolics have been extruded.

Low Pressure Moldings. These are made from asbestos or graphite-filled molding powders in metal molds by curing in an autoclave in hot air under high pressure. Processing equipment for chemical plants made in this way includes cylindrical and horizontal tanks; packed, bubble cap, sieve plate, and cascade towers; pumps; agitators; and flanged pipe. Single pieces weighing up to four tons have been made. These pieces are resistant to a wide range of chemicals. The larger pieces are reinforced by external wooden or metal staves or ribs for greater strength.

Cast Phenolics. These are made in a wide range of attractive colors in transparent, translucent, and opaque types. They are produced in standard shapes such as rods in a variety of cross sections, bars, tubes, and sheets. They are machined into a variety of knobs, handles, game parts (see Figure 6.4), buckles, jewelry, and novelties. Special castings are used for cutlery handles, clock cases, pipe stems, and many similar items.

"Microballoons." Microballoons are hollow spheres of phenolic plastic 0.0002 to 0.0032 inches in diameter, filled with nitrogen. These are used in a blanketing layer of about ½ inch on the surface of crude oil or other volatile liquids to reduce evaporation losses. A ½-inch layer is said to reduce crude oil evaporation losses by 80 to 90 per cent. Microballoons may be mixed with melted phenolic or epoxy resins to produce the so-called *"syntactic"* foams.

Foamed Material. Foamed phenolic products are made by adding a chemical blowing agent to liquid molding resins or by the syntactic method.

Phenolic-Rubber Compounding. Small amounts of rubber added to phenolic molding compounds tend to improve shock and

Fig. 6.4—Cast phenolic game parts. (Courtesy Catalin Corporation)

fatigue resistance. On the other hand, it is possible to plasticize, reinforce, and harden rubber compounds by adding phenolics to them.

Acrylonitrile-butadiene copolymers show best compatibility with phenolic resins, particularly those with medium to high acrylonitrile content. Low rubber content compositions are used in bowling balls and in electrical and mechanical parts subject to considerable vibration and shock. When nearly equal amounts of the rubber and phenolic resin are used, the product is suitable for steam valve disks and storage battery cases. High rubber content (75 to 90 per cent) compounds are used for shoe soles.

Styrene-butadiene copolymers are not as compatible as the acrylonitrile rubbers with phenolics, but have some use in blends. Phenolic resins increase the hardness of neoprene. Phenolic resins act as curing agents for butyl rubber. These rubbers are said to have better resistance to air and steam at high temperatures than the rubber vulcanized with sulfur. Natural rubber and phenolics blends are similar to those from styrene rubbers.

Trademark names: Mouldrite, Plyophen, Resinox, Synvarite

UREA-FORMALDEHYDE RESINS

Production

Urea and formaldehyde react under alkaline conditions to produce either monomethylol or dimethylol urea. A simplified form of the reaction to form the monomethylol compound is as follows:

$$NH_2 \cdot CO \cdot NH_2 + CH_2O \longrightarrow NH_2 \cdot CO \cdot NHCH_2OH$$

Continued reaction results in chains which then cross-link into a thermally set compound.

Aqueous solutions of urea and formaldehyde with an alkaline catalyst such as calcium, sodium, or ammonium hydroxide are reacted together in nickel-clad, stainless steel, or glass-lined reactors. The reaction is stopped by the addition of acid; and a filler, either alpha cellulose or wood flour, is added and the water evaporated. When alpha cellulose is used, it is no longer visible in the molded product but is apparently dispersed in a state of molecular division, or perhaps combined chemically.

Colorants, mold lubricants, and acid catalyzers are added to the molding powder. If acids are added, the final cross-linking proceeds rapidly. Usually latent acid catalysts which react only at temperatures above ordinary room temperatures are used. Examples are aniline hydrochloride, dimethyl oxalate, and ammonium chloride. A small amount of urea or thiourea is also added to react with any excess formaldehyde and to form crosslinks.

Applications

The urea resins are readily molded by compression, setting up thermally in the mold. They have good tensile strength and hardness but are inferior to some phenolics in impact strength. Resistance to water, particularly hot water, is inferior to phenolics. They have good resistance to weak alkalies and most organic solvents but not to acids or strong alkalies. Their high dielectric strength and nontracking property make them suitable for many electrical uses. The urea resins are transparent. The alpha cellulose filled material is translucent and can be colored in a wide variety of pastel colors. This, together with light weight and good impact strength compared to glass and most ceramics, makes it suitable for use in electric light shades and reflectors and in decorative panels. During the early development of the ureas the unfilled product was used as a substitute for glass in windows. The tendency to crazing and the generally poor resistance to weathering have prevented any utilization in this field.

Molded ureas and phenolics are competitive for many products. The ureas are somewhat more expensive but have the advantage of being available in a wide range of light colors. Among the products are buttons, knobs, electric switch and outlet plates, grocery and meat scale housings, bottle closures, and clock cases.

The resin is used both as a general adhesive and in laminates. One product with a latent catalyst is packaged dry and needs only mixing with water to produce a thermosetting adhesive which sets up at ordinary room temperatures. Sometimes a cellulose filler is added to control shrinkage. Extenders such as wheat flour may be added to lower the cost. Urea adhesives are used extensively in production of plywood as well as decorative laminates with a paper or cloth base. They are also used for bonding foundry sand cores.

Wood can be impregnated with a water solution of methylol urea which can then be polymerized. This produces a dense, hard, dimensionally stable product. Urea resins are also used for treating paper and textiles. In the treatment of cotton or rayon it is dipped into a water solution of the intermediate resin. The resin apparently reacts with cotton much as with the alpha cellulose filler. When the cloth is dried and heated, cross-linking occurs. This imparts some crease and crush resistance to the cloth together with a reduction in shrinkage. This reduces the amount of ironing necessary after laundering. The resin is added to paper pulp before forming into a sheet; the addition of 2 to 10 per cent increases the wet strength of the paper considerably.

When a higher alcohol, such as butanol, is reacted with the intermediate-stage urea resin, an ether linkage is formed between it and a methylol group. When this reaction product is heated above its melting point, it becomes resinous with the loss of some alcohol. With further heating, some cross-linking occurs. This type of resin is used, usually mixed with alkyd resins, in production of white and light-colored enamel-like coatings used on refrigerators and similar items.

Trademark names: Arodure, Beetle, Sylplast, Tetra-Ria

MELAMINE-FORMALDEHYDE RESINS

Production

Melamine, $(CN)_3(NH_2)_3$, reacts similarly to urea under neutral or slightly alkaline conditions to form methylol melamines, the monomers of the resin. The reaction is carried out in much the same manner as for urea resins.

Applications

Melamine resins, like urea resins, are transparent and can readily be colored in light, pastel shades. They can be filled with alpha cellulose, like ureas, to give translucent products. Other fillers such as asbestos, cotton flock, macerated fabric, and glass fiber result in products which are opaque but which have properties suitable for many other applications. Melamine resins are stronger and have a much lower water absorption than the urea resins. Melamine molded products, in contrast with ureas, can be used above the boiling point of water continuously. A considerable part of the melamine resin production goes into the manufacture of tableware. Melamine tableware (see Figure 6.5), because of its superior resistance to chipping and breakage and its light weight, is preferred to chinaware for use on air-

Fig. 6.5—Dinnerware of Melamine plastic. (Courtesy Lenox Plastics Division, Lenox, Inc.)

planes, naval ships, for children, and in institutions serving large numbers of people. Melamine moldings can be used for electrical parts because they operate at higher temperatures than ureas. They can also be used out-of-doors. In general, molded melamine products are better than similar urea products but more expensive.

Melamine resins, like urea resins, are used in light-colored, laminated products and to add wet strength to paper. Melamine adhesives have excellent resistance to boiling water, but they harden at higher temperatures than the urea products, making them unsuitable for cold-pressing applications. The methyl ether of methylol melamines is used to shrinkproof wool. Melamine coating resins, made similarly to those from urea, are superior to the urea resins.

Trademark name: Cymel, Risemene, Melmac

The phenolic, urea, and melamine plastics, being thermosetting, do not soften upon heating. Phenolic moldings are available only in dark colors such as brown or black, while urea and melamine polymers can be used in light, translucent forms as well as a wide range of colors.

ANILINE-FORMALDEHYDE RESINS

Production

When one mol of aniline ($C_6H_5NH_2$) in acid solution is re-acted with one mol of formaldehyde (HCHO), a soluble and fusible resin is produced. It probably is a chain polymer made up of repeating units with the formula $—C_6H_5NCH_2—$. With a formaldehyde-aniline ratio of 2:1 under strongly acid conditions, cross-linking will occur.

Uses

Most of the commercial resin is the thermoplastic material. It is a reddish-brown, translucent resin with low power factor, high dielectric and mechanical strength, low moisture absorption, and excellent machinability.

It is used for insulation in high frequency electrical work for such applications as coil forms, tube bases, and antenna housings. It also is used in other electrical applications because of its low conductivity in the presence of moisture.

CHAPTER 7

Polyolefins

POLYETHYLENE

Ethylene, $H_2C{=}CH_2$, is readily produced by cracking certain petroleum fractions. It can be polymerized, using oxygen or a peroxide as an initiator or catalyst, into long thermoplastic chains. The early development of the polymer was retarded because of the high pressures necessary and the difficulty in controlling the reaction. In contrast, the substituted ethylenes, such as vinyl chloride and styrene, are readily polymerized at low pressures. Ethylene is currently polymerized by two general methods, usually referred to as "high pressure" and "low pressure" polymerizations.

Production

High Pressure Polymerization. This polymerization is carried out with an oxidizing catalyst or initiator, such as benzoyl peroxide, and requires pressures in the range of 15,000 to 45,000 psi and temperatures from 175° to 570° F. The exothermic nature of the reaction makes efficient well-controlled heat removal necessary.

The heat control may be facilitated by polymerizing in small-bore jacketed tubes, but there is a tendency for the tubes to be blocked by the polymer. Another method is to use an autoclave with efficient agitation combined with countercurrent heating of the incoming ethylene by the polymerizing ethylene. Since the yields are usually between 10 to 30 per cent, the polymer must be separated from the monomer and the latter returned to the reactor.

Low Pressure Polymerization. The "low pressure" polymerizations are carried out at pressures from 45 to 1175 psi and temperatures from 120° to 520° F. The initiators are of the "stereospecific catalyst" type described in Chapter 1. A typical Ziegler catalyst might be produced from aluminum triethyl and titanium tetrachloride. Operating conditions are from about 45 to 60 psi at from about 120° to 160° F. The ethylene is mixed with an unreactive hydrocarbon solvent and the catalyst in a reactor. The catalyst may be destroyed by adding an alcohol and the polymer separated by filtration. Another procedure involves heating the mixture to dissolve the polymer, decomposing the catalyst with water, and cooling to separate the polymer. In the Phillips Petroleum Company process the chromium trioxide-promoted silica-alumina catalyst is suspended in a saturated hydrocarbon used as solvent. The catalyst slurry is mixed with the ethylene in an agitated reactor at 260° to 350° F and 250 to 500 psi. After flashing out any unreacted ethylene, the polymer solution is centrifuged or filtered to remove the catalyst. The solvent may be flashed with steam or the polymer precipitated by cooling. The Standard Oil Company method uses a promoted metal oxide catalyst on a porous support at around 445° to 520° F and 590 to 1175 psi. The reactor is equipped with an agitator and cooling coils or jacket. The catalyst is filtered from the solution of the polymer.

Low Molecular Weight Polymerization. Polymers with molecular weights less than 10,000 are usually considered as low molecular weight products. They may be produced by polymerizing ethylene at 140° to 325° F and 1500 to 4400 psi in the presence of an inert solvent such as benzene with a peroxide type catalyst such as benzoyl peroxide.

Properties

In general, polyethylene may be said to fall between rigid and nonrigid plastics in flexibility (see Figure 7.1). *Low density*

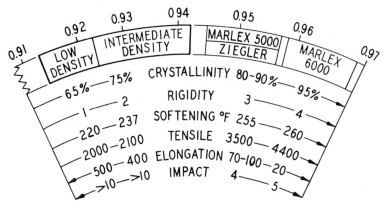

Fig. 7.1—Polyethylene density range. (Courtesy Phillips Petroleum Company)

(0.910 to 0.925 g per cc) resins are made by the older high pressure method; *high density* (0.942 to 0.965 g per cc) by the newer low pressure methods. An intermediate *medium density* is also recognized. These designations are arbitrary since there is a gradual change of properties over a rather wide range.

The branched chain structure of the older high pressure types resulted in low densities, melting points, crystallinity, stiffness, and tensile strength. The unbranched chain structure of the low pressure types results in higher values for these properties. Usually, increasing the molecular weight increases impact strength, low temperature toughness, and melt viscosity. Low molecular weight polyethylenes are waxlike.

In general, polyethylene has excellent chemical resistance. At room temperature it is insoluble in most organic solvents and is resistant to other chemicals such as inorganic acids, alcohols, and fatty oils. Polyethylene shows some oxidation by air at higher temperatures or on exposure to ultraviolet light. Carbon black is an effective antioxidant. Electrical properties are excellent.

Applications

Low molecular weight polyethylene is added to wax used for coatings to increase scuff and grease resistance and to produce optimum sealing temperatures. It is also used in inks, paints, rubber compounds, and polishes.

The higher molecular weight polyethylenes are used in the manufacture of a wide variety of products by such techniques as injection molding, blow-molding, extrusion, thermoforming, and welding. Products include a variety of housewares (see Figure 7.2), electrical insulating parts, pump parts, chemical laboratory ware, pipes and ropes, toys, coatings, and many applications of a miscellaneous nature. Large numbers of bottles, both rigid and squeeze, are blow-molded. Drums, drum liners, and similar pieces up to 50-gallon capacity have been blow-molded. Polyethylene is widely used as a wire and cable covering.

Cross-linked Polyethylene

Irradiated Polyethylene. Although various types of radiation may be used to modify polyethylene, electron beam bombardment using beam voltages of the order one million is commonly used for industrial radiation. Irradiating causes cross-linking, resulting in an increase in mechanical strength, softening tempera-

Fig. 7.2—Typical polyethylene products. (Courtesy Phillips Petroleum Company)

ture, and chemical resistance. The difficulty in securing deep penetration limits irradiation to films, tubes, sheets, and thin sections of molded objects. The tape is used for winding electrical coils and the film has some packaging applications.

Chemically Cross-linked Polyethylene. A thermosetting product can also be produced by adding a suitable peroxide and "curing" under heat and pressure. The peroxide splits off hydrogen from the polyethylene chain, forming activated sites which cross-link. It is also possible to add as much as 300 per cent carbon black which cross-links with the polyethylene. The cross-linked material, although thermosetting, can be extruded, transfer or compression molded, and with proper care, injection molded. It has better strength than ordinary polyethylene and good weather resistance. Principal uses are in cable insulation and in pipe for handling chemical solutions.

Ethylene-Butene Copolymer

The copolymer of ethylene and 1-butene has greater resistance to environmental stress-cracking than the high density polyethylenes and is used in blow-molded bottles for detergent packaging and as a wire and cable coating. It can also be injection molded and made into fibers and high clarity films.

Ionomers

The term "ionomer" was coined by the Du Pont Company "to describe a new polymer which contains both organic and inorganic materials linked by both covalent bonds and ionic bonds." The major constituent of the product under development is ethylene. The ionic bonds are said to involve anions from the polyethylene chain and metallic cations such as sodium, potassium, magnesium, or zinc.

The properties are given as high transparency, toughness, unusual resilience, and resistance to oils and greases. The ionomer can be processed by the usual thermoplastic techniques. Potential uses are many, including vacuum packagings, bottles, houseware, goggles, shields, refrigerator trays, automotive steering wheels and trim, toys, novelties, and electrical parts.

Trademark names: Alathon, Alkathene, Ampacet, Dylan, Ethafoam, Fortiflex, Hi-fax, Marlex, Petrothene, Poly-Eth, Surlyn (ionomer)

POLYPROPYLENE

Production

Polypropylene is produced in a similar manner to high density polyethylene by either batch or continuous methods using stereospecific catalysts. Pressures range up to 600 psi and temperatures from 70° to 250° F. Either the isotactic or atactic chain type can be produced, although the former with its higher crystallinity is the commoner.

Properties

Polypropylene has the lowest specific gravity of any plastic. It has higher melting points, compression strengths, and flexural strengths than high density polyethylene. Tensile strength averages a little better. Stiffness and hardness at ordinary temperatures are good. It has about the same excellent resistance to chemicals as polyethylene. It has an exceptional high flex life. It does not stress crack.

Polypropylene can be injection molded, thermoformed, and extruded. In powdered form it can be used for fluidized-bed coating, sintering, and rotational casting. It is readily welded.

Applications

Polypropylene is competitive in many applications with polyethylene (see Figure 7.3). Because of its higher melting point (about 335° F) it is preferred over polyethylene for some applications such as hot water pipe and hospital items (see Figure 7.4) requiring heat sterilization. It is injection molded into relatively large parts such as wastepaper baskets, tote boxes, tubs for washing machines, and toilet seat hinges (see Figure 7.5).

Polypropylene is used in items incorporating integral hinges, such as key cases, lunch boxes, loose-leaf binders, and automobile accelerator pedals. Tests on these hinges have shown flex cycles of many millions without breakage.

In addition to pipe, tubing, and various profiles, polypropylene is extruded into fibers, sheets, films, and wire and cable

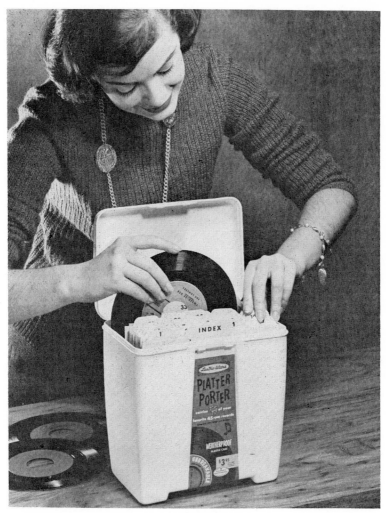

Fig. 7.3—Record holder of polypropylene. (Courtesy Hercules Powder Company)

coverings. The fibers are made into textile items. Carpets show long-wear life and resistance to weathering, making them suitable for outdoor and automobile use. The fibers are also being used in blends with other fibers in a variety of clothing.

Glass- and asbestos-reinforced polypropylenes are used in applications requiring greater stiffness and heat resistance.

Trademark names: Escon, Moplen, Olemer, Pro-fax, Marlex

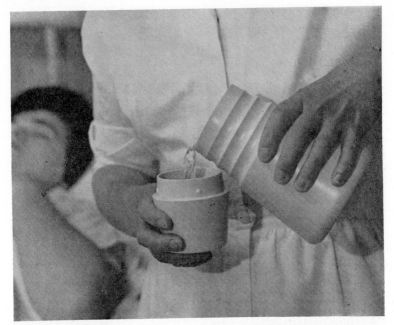

Fig. 7.4—Polypropylene hospital carafe. (Courtesy Hercules Powder
Company)

Polyallomers

Polyallomers are highly crystalline copolymers of propylene
and ethylene. They have many of the properties of both high
density polyethylene and polypropylene. They have excellent
impact strength, resistance to abrasion, stress-crack resistance,
and hinge characteristics. They are easily formed by injection
molding, extrusion, and thermoforming. They have been used
in injection-molded objects such as typewriter cases and fishing-
tackle boxes where their hinge properties are utilized. Vacuum-
formed sheets have been used in embossed luggage shells.

Other Copolymers

Copolymers of ethylene with vinyl acetate (EVA) and ethy-
lene-ethyl acrylates (EEA) are available. In general, the polymers
low in the acetate or acrylate do not vary much in properties from
polyethylene. Those with increasing amounts tend to develop

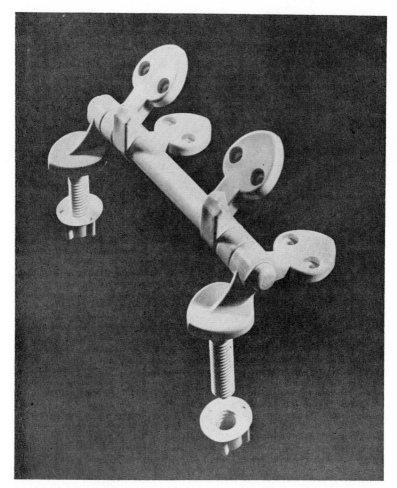

Fig. 7.5—Polypropylene hinge. (Courtesy Hercules Powder Company)

elastomeric properties. Tensile strength and stiffness are less for the copolymers. They can be compounded with fillers and stand weathering better than polyethylene. They can be formed by the same processes used for polyethylene. They are used in products where somewhat greater flexibility than polyethylene is required. These uses include flexible tubing, squeeze toys, film and sheeting, wire and cable covering, and medicine droppers.

A newer copolymer is that of ethylene-*n*-vinyl carbazole. This is very stable at temperatures up to 465° F and has very good low temperature flexibility. It is highly resistant to stress cracking. Suggested uses include hot and cold water and gas pipes, wire and cable coverings, and gasoline tanks.

FLUOROPLASTICS

Polytetrafluoroethylene

Properties. Tetrafluoroethylene ($F_2C{=}CF_2$) may be considered as ethylene in which fluorine has been substituted for all of the hydrogen. It is polymerized in the presence of water and a peroxide catalyst in a stainless steel autoclave. The polymer (PTFE) may be produced in either a granular or dispersion form.

PTFE is completely resistant to all chemical attack except by molten sodium, fluorine, and some fluorine compounds. It may be used in a temperature range of $-450°$ F to $550°$ F. It has a very low coefficient of friction and excellent antistick properties and electrical properties.

Applications. Although classed as a thermoplastic, PTFE cannot be molded by conventional methods. Techniques similar to those used in powder metallurgy, involving preforming at high pressures followed by sintering, are used. It is also possible to extrude it in the form of rods, tubes, and fibers by using an organic extrusion aid which is later vaporized off and the product sintered. PTFE dispersions may be applied to surfaces by dipping or spraying, followed by drying and sintering. Films may be cast from the dispersions. Tape is made by shaving from molded pieces or by extrusion.

It is used in electrical insulators, wire covering, valve seats, seals, and gaskets. It is used in thin coatings as a nonstick surface for objects such as cookware. It is mixed with fillers such as glass, carbon, and metals for use where greater wear might otherwise occur, as in bearings.

Polychlorotrifluoroethylene

This polymer (CTFE or CFE) has chlorine substituted for one fluorine of the original monomer. It is similar in properties to PTFE, although somewhat less resistant to heat and chemicals. It does not have the low friction properties of PTFE. It can be formed by compression, injection, and transfer molding, and by extrusion.

CTFE is used largely for electrical insulation and for valves, gaskets, and seals. It is particularly used at low temperatures.

Tetrafluoroethylene and Hexafluoropropylene Copolymer

This copolymer (FEP) is similar to PTFE with somewhat less chemical resistance and lower melting point. It has similar electrical properties. It can be injection molded, extruded, and vacuum-formed. It may be glass-filled. It is used for laboratory ware, bottles, wire insulation, and corrosion-resistant pipe lining. It is also available in films and fibers.

Other Fluoroplastics

Vinyl fluoride and vinylidene fluoride are discussed under Vinyl Plastics (Chapter 8). The copolymer of vinylidene fluoride and hexafluoropropylene, which has elastomeric properties, is described under Rubbers (Chapter 14).

Trademark names: Halon (a copolymer), Kel-F (CTFE), Rulon (TFE), Teflon (TFE and FEP)

Comparative Physical Properties

Polyethylene moldings usually have a somewhat greasy feel and less rigidity than phenolics. They soften at boiling-water temperature and will float on water. Polypropylene is similar but has a higher melting point and greater rigidity. Polytetrafluoroethylene has a greasy feel but greater rigidity. It decomposes upon heating above 550° F without melting.

CHAPTER 8

Vinyl Plastics

Vinyl compounds have the general formula $H_2C\!=\!CHX$, where X may be hydrogen, an alkyl group, an aryl group, or a negative atom or group, such as halogen, hydroxy, or acetate. The polymers of vinyl chloride, vinyl acetate, and vinylidene chloride are commonly thought of as vinyls although polyacrylates, polystyrene, polyethylene, and related polymers, discussed in other chapters, may also be considered as vinyls. With the exception of ethylene these monomers can be readily polymerized using an organic peroxide as a catalyst or initiator. Apparently the greater polarity of the substituted compounds makes them more reactive than ethylene.

VINYL CHLORIDE AND ACETATE POLYMERS

Production

Vinyl chloride may be produced from acetylene by the following reaction:

$$C_2H_2 + HCl \longrightarrow C_2H_3Cl$$

When ethylene is used as a raw material, a two-step synthesis is commonly used:

$$C_2H_4 + Cl_2 \longrightarrow C_2H_4Cl_2$$
$$C_2H_4Cl_2 \longrightarrow C_2H_3Cl + HCl$$

A new process uses oxychlorination of ethylene to produce dichloroethane:

$$C_2H_4 + \tfrac{1}{2}\, O_2 + 2\ HCl \longrightarrow C_2H_4Cl_2 + H_2O$$

The dichloroethane is then dehydrochlorinated conventionally to produce the vinyl chloride.

While both vinyl chloride and vinyl acetate can be polymerized by all four common methods, mass polymerization is used but little. Some solvent polymerization is carried out. Emulsion polymerization using an organic oxidizing agent as a catalyst and a small amount of a reducing agent to form a redox system is commonly used. Since much of the vinyl acetate polymer is sold for application in the emulsion form, the emulsion is frequently spray-dried without washing out the emulsifier. Considerable quantities of both polymers are produced by suspension polymerization.

A new method for polymerizing vinyl chloride is a two-stage mass system using azodiisobutyronitrile as an initiator. About 10 per cent of the monomer is polymerized in a vertical reactor or prepolymerizer using high speed agitation. It is then pumped to a horizontal reactor agitated by a slow speed ribbon blender. From 75 to 80 per cent of the monomer is polymerized, the remainder being evaporated off and recovered.

Properties

Unplasticized polyvinyl chloride[1] is a hard, tough material with a density of 1.4. It has excellent resistance to many chemicals, but is not recommended for use with acetone, ketones, ethers, and aromatic and chlorinated compounds. For example, cyclohexanone, nitrobenzene, and tetrahydrofuran are solvents for polyvinyl chloride at room temperatures. It has excellent resistance at room temperatures to other organic chemicals such as most alcohols, hexane, linseed oil, mineral oils, and phenol, as well as many inorganic chemical solutions.

Since PVC tends to break down under molding temperatures and form free hydrogen chloride, and since it is also affected by sunlight, various stabilizers are added to prevent deterioration.

---

[1] Sometimes written poly(vinyl chloride) to indicate that the prefix "poly" refers to "vinyl chloride" not simply "vinyl." It is conveniently abbreviated to PVC.

These include alkaline oxides, hydroxides, carbonates, amines, various lead salts, barium-cadmium soap, and esters such as dibutyl tin dilaurate.

The original high molecular weight PVC has been modified to produce a lower molecular weight material which is easier to process but is lower in strength and chemical resistance. Copolymerizing the chloride with 5 to 15 per cent of the acetate produces a material with properties between those of the chloride and acetate polymers which is superior in many respects to both. The copolymers are more readily molded than the PVC, have good strength, and low water absorption. Resistance to chemicals is somewhat poorer. Strengths and melting points may be modified by varying the molecular weight. Another method of modifying is by the addition of plasticizers. This makes possible a material ranging from the hard, strong molding product to flexible vinyl sheet and film.

Unplasticized Products

Unplasticized polyvinyl chloride and the copolymer may be processed by extrusion, calendering, compression, transfer, and injection molding at temperatures between 325° and 400° F. Common products are phonograph records, printing plates, pencil barrels, toothbrush handles, cosmetic containers, ice-cube trays, and bottles. Siding and eaves spouting for buildings are new products (see Figure 8.1). Among industrial items are pipe fittings, valve parts, and similar items for use with corrosive liquids. Pipe is extruded. Sheets are fabricated by mechanical operations, welding, and thermoforming.

Until recently, PVC was not considered practical for use above 160° F because of rapid decrease in strength. A new normal-impact product is said to retain tensile strength of 3000 psi at 212° F and the new high-impact material a tensile strength of 2000 psi. Previous values of 2300 psi and 2000 psi at 180° F were considered excellent.

Plasticized Products

Plasticized polyvinyl chloride has extensive use. Various liquid plasticizers, such as the octyl, nonyl, and decyl phthalates are used. Phosphates may be used to reduce flammability. Polyesters are useful if migration is a problem. Powdered inert

Fig. 8.1—Rigid polyvinyl chloride rainwater system. (Courtesy Bird & Son, Inc., and B. F. Goodrich Chemical Company)

fillers, lubricants, and color pigments are usually added. Amounts of plasticizer commonly vary from 30 to 80 per cent.

Plasticized polyvinyl chloride can be injection molded, extruded, and formed into sheets or films by calendering. The properties of the plasticized product vary considerably, making it suitable for such products as rainwear, shower curtains, table covers, draperies, packaging, refrigerator bowl covers, baby pants, auto seat covers, belts, handbags, suspenders, wallets, upholstery for furniture, and floor coverings. Extruded products include tubing, hose, weather stripping, and wire insulation. Molded products include electrical insulating parts, gaskets, flashlight lenses, toys, buttons, and grommets.

Organosol, Plastisol, and Emulsion Products

In the form of organosol and plastisol dispersions, polyvinyl chloride is made into cast films, coated fabrics, coated papers, and foams. Metal dish racks are coated with plastisols or organosols by dipping. Items such as gloves and rain boots are also made by dipping. Dolls, squeeze tubes, and toys are made by slush or centrifugal molding of plastisols. Plastigels are used in caulking compounds, gaskets, and hand-molding compounds.

Polyvinyl chloride latex is also used directly in the manufacture of sheets, coatings, and elastomeric items. This eliminates several steps otherwise necessary in manufacturing these products. The techniques involved are essentially those used with natural rubber latex.

Polyvinyl acetate emulsions are available in various particle sizes, viscosities, and emulsifying systems. Alone, or blended with other adhesives, they are used extensively as adhesives. Mixtures of paper pulp, wood flour, cork, and similar materials with the emulsions may be molded with heat and pressure. The emulsions are also used in coatings of various types.

Propylene Modified PVC

Less than 25 per cent propylene added during polymerization produces a resin which may be extruded at lower temperatures than the usual PVC. It is said to be better for blow-molding bottles since stabilizers which can be used at the lower temperature make possible a transparent product.

Polyvinyl Dichloride

Polyvinyl dichloride (PVDC) is very similar in properties and applications to PVC but will withstand service temperatures 40° to 60° F higher. It is also self-extinguishing and has excellent electrical properties.

Trademark names: Blacar, Elvax, Exon, Geon, Insular, Marvinol, Opalon, Pliovac, Vygen, Vyram

VINYL FLUORIDE POLYMERS

Polyvinyl Fluoride

The polymer of vinyl fluoride (PVF) is available in the film form. It has high abrasion resistance, chemical inertness, and weatherability. Its principal application has been as a laminate for architectural facings.

Polyvinylidene Fluoride

Although this polymer has excellent chemical resistance, it is lower in chemical resistance, electrical properties, temperature range, and antistick properties than TFE. It can be processed by extrusion and injection and compression molding. Among its applications are valves, impellers, electrical insulation, and chemical tubing.

Trademark names: Kynar $(CH_2{=}CF_2)_n$, Tedlar $(PVF)_n$

POLYVINYL ALCOHOL AND THE ACETALS

Production

Since vinyl alcohol is unstable, polyvinyl alcohol, $(H_2C{=}CHOH)_n$, is produced by reacting polyvinyl acetate with either ethanol or methanol. The acetals are produced by reacting an aldehyde with polyvinyl alcohol using an acid catalyst. Three acetals are produced: polyvinyl acetal, made with acetaldehyde; polyvinyl butyral, made with butyraldehyde; and polyvinyl formal, made with formaldehyde.

Properties and Uses

The properties of polyvinyl alcohol are determined by the extent of the "hydrolysis" (actually alcoholysis) and the molecular weight of the acetate. Completely "hydrolyzed" grades, having between 1 and 15 per cent residual acetate groups, swell in cold water and are completely soluble in hot water. They are insoluble in many organic compounds, including the chlorinated hydrocarbons. If the alcoholysis is less than 50 per cent the product is not soluble in water, vegetable oils, and petroleum fractions, but soluble in many organic solvents. Water sensitivity, tensile strength, tear resistance, elongation, and flexibility increase with the decrease in molecular weight.

Plasticized polyvinyl alcohol can be compression molded and extruded. Common plasticizers are glycerin, ethylene glycol, and triethylene glycol. The finished products are well suited for applications not requiring a high degree of water resistance but requiring a high degree of resistance to organic solvents and greases. The tubing is used in lubricating equipment and for handling paints, lacquers, dry cleaning solvents, fire extinguisher fluids, and refrigerants. The film, which is transparent, strong, and resistant to tearing, is used for bags and protective clothing for laboratory and plant use. Polyvinyl alcohol coatings applied from water solutions containing ammonium dichromate become water insoluble when exposed to ultraviolet light. This is the basis for its use in stencil screens and photolithographic printing plates. Polyvinyl alcohol can be reacted with acrylonitrile to form a rubbery resin or to form a graft polymer suitable for fibers.

Trademark names: Elvanol, Gelvatol

Polyvinyl acetal is produced in several grades of varying solubility in organic solvents. The largest use is in adhesives, impregnating, coatings, photographic film, sheets, rods, and tubes. *Polyvinyl butyral* is tough, flexible, and shock resistant over a wide temperature range, with excellent adhesion for glass, metal, phenolic resins, and cellulosic materials. It is the standard material for safety glass lamination and is used in wash primers for metal finishing. *Polyvinyl formal* is a tough resin which can be molded, extruded, and cast. It is water soluble but completely resistant to oils, gasoline, fats, waxes, and alkalies. It is used as a wash primer for metals, as an adhesive, and in blends with phenolic resins.

Trademark names: Butacite (acetate), Butvar (butyral), Formvar (formal)

SARAN

Production

Vinylidene chloride ($H_2C{=}CCl_2$) polymerizes readily in the presence of peroxides. While it can be mass polymerized, emulsion polymerization using redox technique allows better control and is a more satisfactory method. Because of difficulties in plasticizing and molding the pure polymer, the commercial product, known as *Saran,* is a copolymer with either vinyl chloride or acrylonitrile, ranging upward from 73 per cent vinylidene chloride.

Properties and Uses

The softening range of Saran varies from about 160° to 350° F. It is commonly formed by extrusion and injection mold-

ing. When cooled rapidly it is amorphous, soft, weak, and plia-
ble. Cooling in the mold at about 200° F or cold drawing develops a crystalline structure.

Saran has good resistance to most organic solvents and to
common acids and alkalies. It is nonflammable and has good mechanical properties, toughness, and durability. It has good stability to aging and is tasteless, orderless, and nontoxic. Saran films
show very low water vapor transmission rates and remain flexible
at low temperatures.

Saran tubing is used in laboratories and plants to handle
solvents and chemical solutions. Rigid saran and saran-lined
steel pipe are used industrially. Saran is also used in fibers
largely for outdoor upholstery and automobile seat covers. Saran
film is used in the food industry.

CHAPTER 9

Styrene and Acrylic Plastics

POLYSTYRENE

Styrene (vinyl benzene) ($C_6H_5CH{=}CH_2$), an aromatic vinyl compound, polymerizes by addition readily, even without a catalyst. An organic peroxide, such as benzoyl peroxide, either alone or in a redox combination, will speed up the polymerization rate.

Production

Styrene has been polymerized by all four methods: mass, solution, emulsion, and dispersion. It has been mass polymerized by passing the monomer down a heated tower and extruding the polymer out at the bottom. This product had a low molecular weight. By partially polymerizing the monomer in kettles ahead of the tower it was possible to produce a polymer with a molecular weight up to 150,000. By working the partially polymerized product on steam-heated rolls while evaporating off the monomer under vacuum, it is possible to raise the molecular weight still further to 400,000. In another method, polymerization is secured continuously in a heated horizontal trough with a conveyor screw. A batch polymerization process uses a modified plate and frame filter press to produce 200-pound polymer blocks.

Molding powders are produced by emulsion and suspension methods and are usually of low molecular weight. Latices of good stability are made by emulsion polymerization for use as finishes.

Properties and Uses

Polystyrene is transparent and can be dyed or pigmented to produce both transparent and opaque colors. It is thermoplastic and readily injection molded and extruded. It is rigid and brittle, with a characteristic metallic sound when dropped on a hard surface. Because of its high index or refraction it has the property of transmitting light through curved sections. Use is made of this in advertising novelties, in lighting automotive and similar panels, and in lighting signs.

Polystyrene is highly resistant to acids and various water soluble chemicals but is attacked by oils and most organic solvents. It is an excellent electrical insulator. It has a softening point slightly above boiling-water temperature, which limits its use somewhat.

Polystyrene is used in such items as acid bottle closures, refrigerator dishes, costume jewelry, wall tile (see Figure 9.1), electrical parts, flowerpots (see Figure 9.2), and lenses. Expanded polystyrene is used in heat insulation, floats, and packaging.

Glass-reinforced polystyrenes with 20 to 40 per cent glass fibers show increases of 50 per cent in strength and 100 per cent in rigidity as well as considerable improvement in impact strength. These have replaced zinc and aluminum die castings in such products as automotive assemblies, business machines, and appliance housings.

Acrylonitrile copolymers have a higher resistance to various solvents, fats, and other organic compounds than polystyrene. They have higher strength and heat resistance, and greater resistance to stress cracking, but are higher in cost than the homopolymer.

Acrylonitrile-butadiene-styrene (ABS) resins were originally blends of 60 per cent or more styrene-acrylonitrile copolymer with butadiene-acrylonitrile rubber. They are now commonly copolymers of the three monomers. The properties of ABS products can be varied considerably by changing both the relative amounts of the monomers and the manner in which the monomers are attached to each other in the polymer structure. As a group they

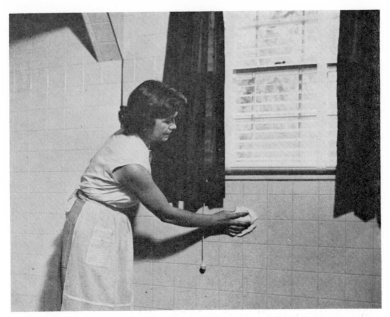

Fig. 9.1—Polystyrene wall tile. (Courtesy Dow Chemical Company)

Fig. 9.2—Injection molding. A polystyrene urn planter. (Courtesy National Automatic Tool Company)

are characterized by improved shock resistance and increased elongation, with good electrical and mechanical properties.

The ABS polymers are processed by injection molding, extrusion, blow-molding, and calendering. They are used in housings of various types where rigidity, strength, and glossy surface are important. In electroplated form they are used in various automotive, appliance, hardware, and housewares applications. Other applications include golf club heads, fan blades, battery cases, bobbins, wheels, and pipe.

Trademark names: **Polystyrene**—Fostalite, Fostarene, Lustrex, Santofome, Styron, Tyril, Zerlon. **ABS**—Cycolac, Cycolon, Kralastic

ACRYLIC RESINS

The acrylic resins are polymers of the esters of acrylic (propenoic) acid ($H_2C{=}CH$ COOH) and methacrylic (alpha methyl acrylic acid) [$H_2C{=}C(CH_3)COOH$]. Acrylic plastic is commonly regarded as that containing 90 per cent methyl methacrylate copolymerized with an ester of acrylic or methacrylic acid. Modified types are made by copolymerizing or blending with other acrylic or with nonacrylic monomers.

Production

A peroxide type catalyst is commonly used for polymerization. All four types of polymerization are used, depending on the type of resin and the end use. Bulk or mass polymerization is usually limited to cast sheets or rods because of difficulties in shrinkage, bubble removal, and heat control. Sheets are cast between glass plates. Rods are made by layer or zone polymerization in a metal tube.

Solution polymerization is carried out in a solvent such as toluene, benzene, or ethyl acetate in a resin kettle. *Emulsion polymerization* usually is in a stainless steel kettle using a redox catalyst system. *Suspension polymerization* uses benzoyl peroxide catalyst with a protective colloid such as bentonite, starch, sodium methacrylate, or magnesium silicate.

Properties

Acrylic resins are transparent, with excellent clarity and high index of refraction. They are not attacked by nonoxidizing acids, weak alkalies, food oils, and common petroleum lubricants. They are attacked by lower alcohols, phenols, chlorinated hydrocarbons and some esters, and ketones. They show excellent weatherability. Electrical properties are good.

Applications

Acrylics may be cast, extruded, injection molded, and dough molded. The sheets may be machined, thermoformed, cemented, and welded.

Acrylic sheets are used in such products as canopies, windows, instrument panels, illuminated signs, lighting fixtures, and shower enclosures. Molded products include automotive parts, knobs, dials, combs, brush backs (see Figure 9.3), and faucet handles (see Figure 9.4). Acrylic polymers are used in dentures and in the embedment of biological specimens.

A copolymer of about 95 per cent ethyl acrylate and 5 per cent chlorethyl vinyl ether is used as an elastomer. An acrylic polymer dispersion forms the base of a new outside house paint said to be superior to oil paints.

Acrylonitrile ($H_2C{=}CHCN$), polymerized alone or with small amounts of vinyl chloride, is used as a fiber. The polymer of alpha chloromethyl acrylate is a harder and more heat-resistant product than the polymethyl acrylate.

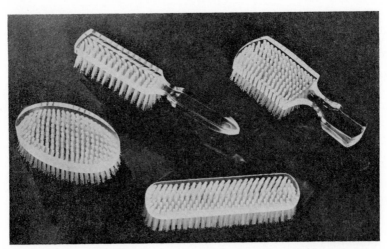

Fig. 9.3—Hairbrush backs of acrylic resin. (Courtesy Rohm and Haas Company)

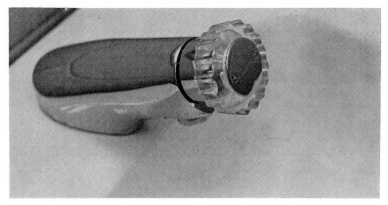

Fig. 9.4—Transparent acrylic faucet handle. (Courtesy Rohm and Haas Company)

Uncolored polystyrene and the acrylic resins are, in general, more rigid and of greater clarity than the polyolefins and ureas. When dropped on a hard surface polystyrene gives off a metallic ring. Acrylic resins have a high index of refraction. The acrylic products stand weathering very well in contrast to the polystyrene.

Trademark names: Acrylite, Lucite, Plexiglas

CYCLIC COMPOUNDS

Coumarone and indene, found in certain coal tar fractions, are polymerized to produce resins which range from those liquid at room temperature to others melting at 300° F. They are soluble in most hydrocarbons. These resins are seldom molded alone but may be used as modifiers or extenders in molding compositions. They are used in varnishes, adhesives, printing inks, and paper coatings.

Trademark name: Cumar

CHAPTER 10

Polyesters and Epoxies

POLYESTERS

Polyesters are produced by reacting polybasic acids and polyhydric alcohols either or both saturated or unsaturated. They may be modified by monobasic acids and monohydric alcohols and may be either thermoplastic chains or thermosetting compounds.

ALKYDS

While the first polyester resin reported was glyceryl tartrate, glyceryl phthalate was the first to be of commercial importance in this country. The reaction between glycerol and phthalic anhydride (see Figure 1.1) occurs at 320° F or above, producing a product which, with continued heating, becomes increasingly viscous until gelation occurs. Upon cooling it solidifies to a hard, brittle, clear, infusible resin of low solubility which originally had a limited use as a bonding agent for mica flakes for electrical insulation work. Later it was found that it could be modified with drying oils or fatty acids to produce a more soluble product, commonly referred to as an *alkyd* resin, suitable for use in coatings.

Production

The reaction is carried out in a stainless steel resin kettle under an inert gas, such as nitrogen or carbon dioxide, by either mass or solution polymerization. The latter method allows better temperature control since heat can be removed in the reflux condenser and the solvent returned to the kettle. The reaction proceeds more slowly, producing a more uniform product.

The modifying oils may be added to the reaction along with the anhydride and the glycerol as free fatty acids or as monoglycerides. One advantage of the fatty acids over oils is that the resin is soluble in the acids and not in the oil. Another advantage is that a more selective control over the properties is possible with the fatty acids than with the oils, which are mixtures of glycerides with two or three different fatty acid radicals to the molecule.

If no modifier is added, the glycerol and phthalic anhydride will cross-link through the third active OH group on the glycerol. If the fatty acid is saturated, cross-linking does not occur and the product is thermoplastic. Other acids or anhydrides and alcohols can be used if at least one is difunctional and one trifunctional. Other resins may also be used as modifiers.

Use

Modified alkyd resins are used extensively in coatings.

UNSATURATED RESINS

Unsaturated resins result from the reaction of polybasic acids and polyhydric alcohols, one of which is unsaturated. Common combinations are either maleic or fumaric acids with a saturated glycol such as ethylene or propylene. Unsaturated polyester resins can be cross-linked with unsaturated monomers such as styrene to produce insoluble, infusible compounds. Since no gaseous by-products are formed, molding can be at relatively low pressures.

Cross-linking with diallyl phthalate produces a somewhat more flexible resin than styrene. Triallyl cyanurate results in greater heat resistance.

Production

Unsaturated resins are manufactured in a similar manner to alkyds. After the acid and alcohol have reacted, the cross-linking agent is added with an inhibitor, such as hydroquinone.

Low Pressure Molding Resins

These resins, which are sometimes called alkyds, are made with mineral fillers such as clay or powdered limestone, and molded in compression molds at pressures from 800 to 1500 psi. They have excellent electrical properties and heat resistance and are used mainly in electrical parts.

The *premix* types contain both the filler and short lengths of reinforcing material such as chopped glass fibers. They are molded at pressures from about 200 to 500 psi and are used mainly in larger electrical parts than the unreinforced materials.

Reinforced Plastics

In these products a catalyst active at 70° to 300° F is usually added a short time before the resin is to be used. Curing schedules may vary from 15 seconds to 30 days, depending upon catalyst and temperature.

The most commonly used reinforcing material is fiber glass used in the form of rovings, reinforcing mats, yarns, surfacing and overlay mats, woven cloths, and woven rovings. Fillers such as calcium carbonate and aluminum silicate are frequently used with the reinforcing materials.

The molding methods used are contact or hand lay-up (see Figure 3.10) molding, vacuum bag molding, pressure bag molding, flexible plunger molding, vacuum injection molding, matched die molding, centrifugal molding, and filament winding. Laminated sheets are formed continuously, using fiber-glass fabric or mat webs which pass through a liquid resin bath. The hand lay-up method is adapted to limited production of products such as boats and car bodies. This method has the advantage of low mold costs and the disadvantage of high labor costs. If a large number of pieces are molded, matched die molding may result in lower overall costs since the lower labor costs may more than offset the die cost per molded unit. This latter cost goes down with increased production.

Glass-reinforced polyesters are used for a large variety of products. Among these are furniture, sports car bodies, truck bodies, trailers (see Figure 10.1), luggage, bathtubs, fishing rods, water and snow skis, dishwasher tubs, telephone booth roofs, and body armor. Boats from small rowboats to motor cruisers (see Figure 10.2) are made from glass-reinforced moldings. Naval landing craft 36 feet long have been molded in one piece. Industrial equipment includes a wide variety of tanks, fume ducts, valve bodies, safety helmets, fans, pumps, and tote boxes.

Fig. 10.1—Camping trailer of glass-reinforced polyester. (Courtesy Knight Manufacturing Company and Reichhold Chemicals, Inc.)

Fig. 10.2—Power boat of glass-reinforced polyester resin. (Courtesy Hatteras Yacht Company and Reichhold Chemicals, Inc.)

Translucent sheets in various colors are used in buildings in the construction of partitions, roofs, awnings, and outside walls. A new development is an imitation stone facing of molded-glass–reinforced resin which is nailed onto the outside of buildings.

Trademark names: Genpol, Glaskyd, Hetron, Laminac, Polylite, Stypol, Vibrin.

Allyl Resins

Diallylphthalate, $C_6H_4(COOCH_2CH{=}CH_2)_2$, in the presence of an alkaline catalyst polymerizes to form a thermally set product. These resins are compounded with various fillers and molded by compression and transfer techniques. They have excellent electrical properties and chemical resistance and can be made in a wide range of colors. Special types may be used up to 500° F. They are used mainly in electrical parts with minor use in more decorative areas.

Trademark name: Dapon

SATURATED RESINS

Fiber- and Film-forming Resins

The common polyester fibers and films are produced from the reaction product of terephthalic acid $C_6H_4(COOH)_2$ and ethylene glycol (CH_2OHCH_2OH). The direct esterification reaction is very slow. The more usual commercial method is to heat the dimethyl ester of the acid with two equivalents of ethylene glycol at about 370° F. After distilling the methanol which is formed, heating is continued to about 530° F to complete the reaction and remove excess glycol. Sometimes the reaction is carried out in a solvent such as naphthalene or diphenyl.

A continuous process is also in use. Ethylene glycol mixed with a zinc-based catalyst flows continuously into the ester interchange reactor. Melted dimethyl terephthalate is also metered in. The by-product ammonia is condensed. Polycondensation of the diglycol terephthalate is completed in a series of three baffled reactors with stirring. Time and temperature (about 530° F) are carefully regulated. The melted polyester can be spun into fiber or extruded into sheets directly from the last reactor.

Trademark name: Mylar (film)

EPOXY RESINS

Production

Typical epoxy resins are made by reacting bisphenol A and epichlorohydrin (see Figure 10.3) in the presence of a caustic material. The molecule consists of the chain of repeating units with terminal epoxy groups at either end. This chain is thermoplastic but it can be cross-linked through the epoxide linkages of the terminal groups. Cross-linking or curing agents include various amines such as ethylenediamine, diethylenetriamine, triethylenetetramine, diethylaminopropylamine, piperidine, m-phenylenediamine, dimethylaminomethyl phenol, and methyl benzyldimethylamine. Certain acid anhydrides are also used, including phthalic anhydride, pyromellitic dianhydride, dodecyl-succinic anhydride, and hexahydrophthalic anhydride.

Since unmodified cured epoxies tend to be brittle, flexibilizing modifiers such as polysulfide rubber are sometimes added. Polyamide resins, made by condensing dimerized or trimerized vegetable oil, such as soybean, are used as curing agents. Unsaturated fatty acids and aryl or alkyl polyamides are used as combined modifiers and curing agents.

Substitution of a phenol-formaldehyde novolac for bisphenol A produces a resin with greater functionality, allowing greater cross-linking, with resulting greater heat resistance. Another group are the cycloaliphatic epoxies produced by reacting cyclic olefine with peracetic acid. In these the epoxy oxygen is on a ring rather than the end of a chain. The condensed structure is said to result in products which are stronger and more heat resistant than those made with bisphenol A. Typical resins are epoxycyclohexyl, epoxycyclopentyl, and epoxydicyclopentyl.

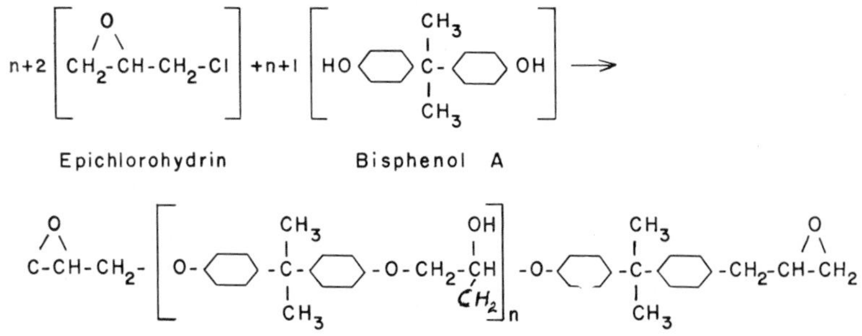

Fig. 10.3—Epoxy reaction.

Uses

Epoxy resins are used with glass-fiber reinforcing for many of the same uses as the glass-reinforced polyesters, being limited mainly by their higher cost. They have better adhesion to glass than polyesters and form products of better strength and chemical resistance. They are used in such items as high pressure pipe, pressure spheres for compressed gases, hydrofoils, storage tanks, exhaust stacks, cover plates on airplane propellers and truck bodies. Both cast and glass fiber laminated epoxies are used in such metal-working tools as draw dies, jigs, and fixtures. They are used as potting compounds for electrical parts, in caulking compounds, and in coatings. Combined with mineral aggregates or fillers they are used over concrete, wooden, or steel floors in food and chemical plants. Epoxies are also used in patching compounds and skid-resistant coatings for highways.

Trademark names: Epi-Rez, Epolite, Epon, Epoxol, Maraset, Tipox, Unox

CHAPTER 11

Polyamides
and Polyurethanes

POLYAMIDES

Two general classes of polyamides are available: nylons and "Versamids." The term *nylon* was defined as: "A generic term for any long chain synthetic polymeric amide which has recurring amide groups as an integral part of the main polymer chain and which is capable of being formed into a filament in which the structural elements are oriented in the direction of the axis." What were originally fiber raw materials are now also important moldable plastics. The Versamids are reaction products of a dimerized fatty acid and a polyamine.

Types of Nylon

Four types of nylon are currently available in this country:

Nylon 6,6—Polyhexamethyleneadipamide:

$$\left[NH(CH_2)_6NHCO(CH_2)_4CO \right]_x$$

Nylon 6—Polycaprolactam:

$$\left[NH(CH_2)_5CO \right]_x$$

Nylon 6,10—Polyhexamethylenesebacamide:

$$\left[\, NH(CH_2)_6NHCO(CH_2)_8CO \,\right]_x$$

Nylon 11—Poly(11-aminoundecanoic acid)

$$\left[\, NH(CH_2)_{10}\ CO \,\right]_x$$

Nylon 12—Poly(12-aminododecanoic acid)

$$\left[\, NH(CH_2)_{11}\ CO \,\right]_x$$

Some other types such as nylons 7, 8, 9, and 12 are in limited production. A new series designated "polycylamides" is being developed. Typical of this series is poly(1,4-cyclohexylenedi-methylenesuberamide).

Production

Nylon 6,6. Hexamethylenediamine is added to a solution of adipic acid until the acid is neutralized. The water solution of the salt is concentrated in an evaporator and then heated in an autoclave to about 500° F. Polymerization is completed under reduced pressure and in an atmosphere of nitrogen to prevent oxidation. The polymer is extruded onto a casting wheel where it is cooled by water to solidify it. Nylon 6,10 is made similarly.

Nylon 6. Both batch and continuous processes of polymerization (see Figure 11.1) are used. One method is to heat the capro-

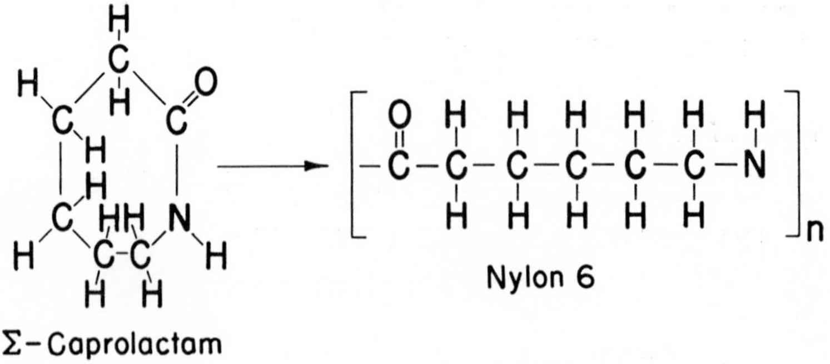

Σ-Caprolactam   Nylon 6

Fig. 11.1—Caprolactam polymerization reaction.

lactam with 10 per cent water and 0.1 per cent acetic acid in a continuous reactor at about 480° F. The acetic acid acts as a chain terminator, thus controlling polymer size. The product from the reactor contains about 10 per cent of the monomer which is separated by leaching with water or distilling under vacuum.

Nylon 11. Polymerization of the 11-aminoundecanoic acid is carried out by heating at about 395° F, during which time by-product water distills off. The acid must be relatively pure to obtain a high molecular weight polymer. The molten product is kept in an atmosphere of inert gas to prevent oxidation. It may be cooled in a manner similar to nylon 6,6 or spun into fibers directly.

Vegetable Oil Products. Dimer and trimer esters formed by heat polymerization of monoesters of fatty acids from soybean or linseed oil are reacted with di- or triamines with stirring at about 212° F until the water is largely distilled off. The condensation is completed by heating between 300° and 460° F.

Properties and Uses

The nylons are thermoplastic and can be molded by injection and extrusion, and at least one variety can be blow-molded. Nylon 6,6 and nylon 6 have been the most important molding nylons in the United States. Nylon 6,6 has greater hardness, abrasion resistance, brittleness, shrinkage, and heat resistance than nylon 6. Nylon 6 has greater impact strength, melt viscosity, and flexibility than nylon 6,6. It is also somewhat more readily molded. The melting point of nylon 6,6, normally about 509° F, can be reduced by the substitution of methyl groups for hydrogens on the carbons. Various copolymers can be prepared by using combinations of different diamines and dibasic acids. In general, these have lower melting points, strengths, and crystallinity than the corresponding homopolymers.

Nylon has excellent resistance to alkalies, petroleum products, and most organic solvents. It is dissolved by hot phenol and formaldehyde and attacked by all but dilute solutions of mineral acids. "Soluble" nylon resin is also soluble in the lower alcohols such as ethyl and methyl.

Because of excellent toughness, resistance to abrasion, low

friction coefficient, and resistance to lubricants, nylon is used widely in small bearings, gears, cams, and other mechanical parts. Because of a combination of excellent mechanical and electrical characteristics it is used in many electrical equipment parts. Some of the automotive applications include speedometer gears, windshield wiper gears, bushings, dome light lenses, fuse holders, voltage regulator bases, and gear couplings. Squirrel cage blower rotors for refrigerators and vacuum cleaners (see Figure 11.2) are now made of nylon. A wide variety of other uses includes water

Fig. 11.2—Nylon vacuum cleaner rotor. (Courtesy E. I. du Pont de Nemours)

hoze nozzles, valve seats, bearing strips, electric shaver casings, clutch facings, football helmets, rifle stocks, racehorse shoes, outboard motor propellers, unbreakable tumblers, combs, and brushes.

Stabilizers may be added to nylon to reduce oxidation at high temperatures and provide better weatherability. As much as 40 per cent glass fibers may be added for greater stiffness and lower creep at higher temperatures. Molybdenum disulfide imparts high lubrication characteristics.

Nylon is used extensively as a textile and bristle fiber material. The soluble type is applied from alcoholic solutions as adhesives, protective coatings, and gaskets. A limited amount of nylon film has been used.

The vegetable oil polyamides have been used in solution as surface coatings and in inks, adhesives, and lacquers. They have been reacted with epoxy resins to form molding and casting compounds. They are used both with epoxy and phenolic resins in coatings.

Modified Nylon

Nylon has been modified in various ways mainly to produce a "nonflat-spotting" fiber for tire cord. One of these (N-44) is reported to be a block copolymer consisting of 75 per cent hexamethylene adipimide and 25 per cent hexamethylene isophthalamide. Another approach has been to spin nylon and a fiber, such as an acrylic, together to form a mixed thread.

A new product contains 50 to 90 per cent nylon and 10 to 50 per cent polyester. The initial commercial product (EF-121) contains 30 per cent nylon 6. Apparently the polyester is completely dispersed in nylon and then melt-spun. It is said that each polymer is present in the fiber as a "discrete entity." This fiber is suitable for use in apparel, carpeting, and upholstery as well as tire cord.

Trademark name: Zytel

POLYURETHANES

Production

The urethane linkage is formed when an isocyanate reacts with a compound containing hydroxyl groups. Polyurethane is produced by the reaction of a diisocyanate and a polyether, polyester, or a hydroxyl-bearing oil such as castor oil. Toluene diisocyanate (TDI) and a polyether such as polyoxypropylene are the most common combination. Other isocyanates used are diphenylmethane diisocyanate (MDI) and polymethylene polyphenyl isocyanate (PAPI).

Applications

Foams. Flexible foams, produced mainly with polyethers, and rigid foams, also mainly using polyethers but some with polyesters, are the major application of polyurethanes. (See Foams in Chapter 15.)

Coatings. Polyurethane coatings are tough, hard, flexible, and abrasion and chemically resistant, but tend to yellow. In greater thicknesses polyurethane is applied as a monolithic floor covering.

Rubber. The elastomeric polyurethane is described under Rubbers, Chapter 14.

Thermoplastic Product. Thermoplastic polyurethanes can be made using linear polyesters or polyethers. These may be injection molded, extruded, and calendered. These are characterized by toughness, abrasion resistance, extended flex life, fuel and solvent resistance, gamma radiation resistance, low-temperature performance, and low gas permeability.

Applications include textile coatings, packaging films, pallet wheels, gears, tank linings, horseshoes, surgical gloves, and guide rollers.

Urethane Latexes. A water-dispersed urethane copolymer is available for application to cloth, paper, and leather. After application it can be fused by heat to characteristic polyurethane properties.

Trademark names: Arothane, Estane, Hetrofoam, Spenkel, Texin

CHAPTER 12

Silicones

Silicones are compounds containing silicon, oxygen, and one or more organic groups. They are usually made up of alternating silicon and oxygen groups with the organic radicals attached to the silicon atoms. The bond energy of the Si—O linkage is about 1.5 times the bond energy of a C—C linkage. When the carbon group is small, such as a methyl group, the Si—C linkage is strong and relatively heat stable. These strong linkages produce stable compounds. The polymers used in plastics are usually produced by the hydrolysis and condensation of intermediate compounds containing halides.

Silicon may be produced first from silica, SiO_2, from sand by reduction with carbon, or the reduction may be carried out together with chlorination in one step. The Grignard reaction may be used as the first step, as for example in the following equation:

$$C_6H_5MgCl + SiCl_4 \longrightarrow C_6H_5SiCl_3 + MgCl_2$$

While this is a useful procedure, a direct reaction with elemental silicon between 300° and 400° C is commonly used commercially as a continuous process for making intermediates such as dimethyldichlorosilane. The typical reaction is:

$$2CH_3Cl + Si \longrightarrow (CH_3)_2SiCl_2$$

The reaction, which is carried out in a tower-like reactor, actually produces several chemicals which must be separated by rectification. For example, a mixture of dimethyldichlorosilane and methyltrichlorosilane separated from the reactor mixture may be hydrolyzed with water and polymerized to form the silicon resins. The typical silicone resin or rubber has a long chain of alternating silicon and oxygen atoms with side chains and terminal groups, such as methyl groups. Linkages between the chains convert the compound into a solid resin or rubber. A great number of these compounds have been produced but only a few are made commercially.

RESINS

Properties and Uses

Silicone resins are used in the form of molding compounds, laminating resins, foams, protective coatings, dipping and impregnating varnishes, and potting materials.

Silicone molding compounds are thermosetting. Inorganic fillers are used. They can be molded by compression or transfer methods and have good molding properties (see Figure 12.1). One of the principal advantages of silicone moldings is their ability to withstand high temperature. Typical flexural strength is 14,000 psi at room temperature and 5,000 psi at 400° F. Electrical properties are very good. Water absorption is low. Silicone molding materials can be molded with inserts and can be machined.

Applications of these molding compounds are largely for products where excellent heat stability and electrical properties together with good strength are needed. These include structural and electrical parts on aircraft and guided missiles. They are used satisfactorily on continuous service up to 600° F and intermittent service up to 1500° F. They are used in electrical parts such as coil bobbins, terminal boards, switches, and fuse assemblies.

Silicone-laminating resins are used in both low- and high-pressure molding. Like the molding compounds, they retain a high percentage of room temperature strength at elevated temperatures. They are used with fiber-glass reinforcing to produce laminates which retain a higher strength-to-weight ratio at high temperatures than many light metals. They have excellent dielectric properties. Typical applications include aircraft ducts, coil forms, transformer spools, terminal boards and strips, and

Fig. 12.1—Typical silicone products. **Left,** molding powder; **above,** pre-form. (Courtesy Dow Corning Corporation)

slot wedges. Silicone resins are also used to bond and laminate such materials as asbestos, mica, and other mineral-reinforcing materials. Flat, rigid laminates may be made with asbestos paper or a combination of asbestos paper and mica splittings. Silicone-treated asbestos paper may be also formed into tubes.

RUBBER

Silicone rubbers are described under Rubbers, Chapter 14.

POTTING MATERIALS

Several silicone potting and encapsulating materials, both modified resins and rubbers, are produced. One variety is a solventless thermosetting resin. This is available as a liquid in two viscosities: 100 and 2,500 centistokes at 77° F. This is cross-linked by a peroxide catalyst, such as dicumyl peroxide. For im-

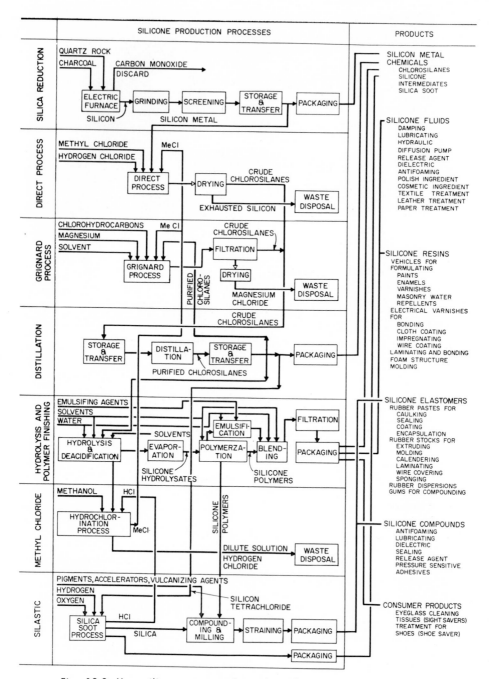

Fig. 12.2—How silicones are made and used. (Courtesy Dow Corning Corporation)

pregnating, no filler is used. For potting, the resin is mixed with a filler such as diatomaceous earth or silica flour. This material will withstand continuous service at 400° F and intermittent duty above 500° F.

Another potting compound known as a *dielectricgel* is a clear liquid which cures, after pouring into place, as a transparent self-healing gel. It has excellent dielectric properties and is stable up to 390° F and down to 76° F.

OTHER SILICONE USES

Silicone products of different compositions are used to coat glass fibers used as reinforcements to secure better bonding, in coatings and varnishes, as lubricants, and antifoaming agents. Silicone production and uses are summarized in Figure 12.2.

Trademark name: Sylgard

CHAPTER 13

Engineering Plastics

The term "engineering plastics" has been applied rather loosely to certain thermoplastic polymers with very good mechanical, electrical, and chemical characteristics which are dimensionally above average in stability, particularly at elevated temperatures. These plastics are suitable for many applications where they replace metals and where they have the advantage of better resistance to chemicals. They are not limited to these applications.

POLYSULFONES

Properties

Polysulfones consist of chains of phenylene rings, ether, and sulfone units. Aryl sulfones have been known as very stable to oxidation and thermal degradation and these properties have been imparted to the chain. The useful temperature range is given as —150° F to over 300° F. Tensile strength and flexural modulus are high, and creep unusually low. Chemical resistance is good to excellent for practically all chemicals except ketones, chlorinated hydrocarbons, aromatics, and other polar organic solvents. Electrical characteristics are very good.

Applications

Polysulfones can be fabricated by extrusion, injection molding, blow-molding, and thermoforming. Applications include switches and other electrical parts, pipe, sheeting, and appliance housing.

POLYPHENYLENE OXIDE

Properties

Polyphenylene oxide (PPO) is made by oxidative coupling of 2,6-dimethylphenol in a solvent, such as o-dichlorobenzene, and a nonsolvent, such as propyl alcohol, with a catalyst consisting of an amine complex of copper salts, and oxygen.

PPO has a brittle point below —275° F and a heat deflection point of about 375° F. It is highly resistant to acids and bases but soluble in certain chlorinated hydrocarbons and aromatic solvents. It has excellent mechanical and electrical properties.

Applications

PPO may be fabricated by injection molding and extrusion. Suggested applications include electrical and electronic products, household appliances, food-handling equipment, pumps, hospital equipment, and plumbing.

POLYALDEHYDES OR POLYACETALS

Production

One type of polyaldehyde is an unbranched polyoxymethylene chain formed by polymerizing formaldehyde in an inert solvent such as hexane with a catalyst such as an amine or cyclic nitrogen-containing compound. Since such a chain with a terminal OH radical would be unstable, terminal methyl or acetyl radicals are inserted instead. Another type is produced by the reacting vaporized trioxane and gaseous catalyst such as boron trifluoride, and cooling. About 2 per cent of a second monomer such as ethylene oxide is copolymerized into the chain for better stability.

Properties and Uses

The polyacetals are thermoplastics and can be formed by injection molding, blow-molding, extrusion, and machining. They have good tensile strength and creep resistance over a considerable range of temperature and humidity and withstand continuous flexural loading. They have good electrical properties and excellent chemical resistance.

The acetals are characterized as "engineering plastics" which can be used to replace die-cast and stamped metals. Applications include high-speed gears, pump impellers, truck bearings, automobile instrument cluster housings, showerheads, aerosol containers, automotive window cranks, and refrigerator door handles.

Trademark names: Celcon, Delrin

POLYCARBONATES

Production

Organic polycarbonates are formed by a reaction between a bifunctional alcohol or phenol and phosgene ($COCl_2$) (see Figure 13.1) or an alkyl or arylcarbonate to give a compound with a carbonate linkage joining the organic units. One commercial product is made by reacting bisphenol A with carbonyl chloride.

Properties and Uses

Polycarbonates are thermoplastic and can be formed by injection molding, extrusion, vacuum molding, welding, and machining. They have a high impact strength and very good heat and creep resistance. They are transparent. They have good chemical resistance, electrical properties, and weathering resistance and are "self-extinguishing." They also fall in the category of "engineering plastics."

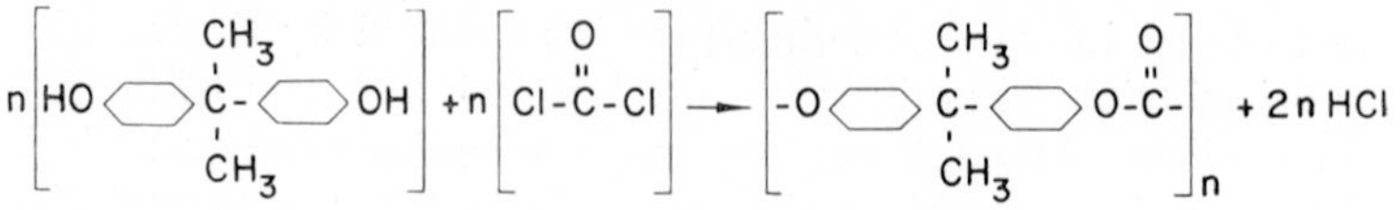

Fig. 13.1—Polycarbonate reaction.

Typical polycarbonate parts include coil forms (see Figure 13.2), pump impellers, battery connectors (see Figure 13.3), cooling fans, marine propellers, and structural housings for power tools. Other applications include safety helmets, heat resistant lenses, and drafting film.

Trademark names: Lexan, Merlon

CHLORINATED POLYETHER

Production

Pentaerythritol is chlorinated and the resulting 3,3-bis(chloromethyl) oxetane (see Figure 13.4) is polymerized to produce a linear, crystalline polyether containing about 46 per cent chlorine. The average molecular weight is about 300,000.

Properties and Uses

Chlorinated polyether is thermoplastic with a melting point of 358° F and a low melt viscosity, allowing it to be readily formed by injection molding or extrusion into strain-free, close-tolerance, dimensionally stable final products. It has low creep

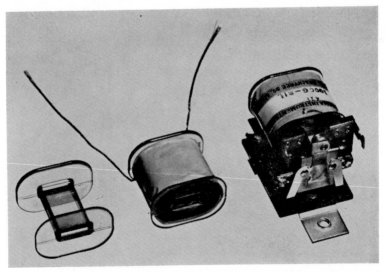

Fig. 13.2—Polycarbonate coil forms. (Courtesy General Electric Company)

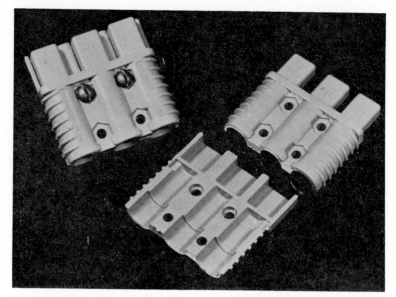

Fig. 13.3—Polycarbonate battery connectors. (Courtesy General Electric Company)

characteristics. It can be used up to 275° F in contact with many chemicals. Certain fillers such as graphite, glass, asbestos, silica, and polymolybdenum disulfide are used or are under study to improve certain properties such as mold shrinkage and thermal expansion. Thermal conductivity is low and electrical properties are good.

It is used largely where chemical resistance is important. These uses include injection-molded valves and pipe fittings; extruded pipe and tubing; and extruded sheets for lining tanks and other chemical equipment. Chlorinated polyether may be also applied as a coating to metal surfaces by the fluidized-bed

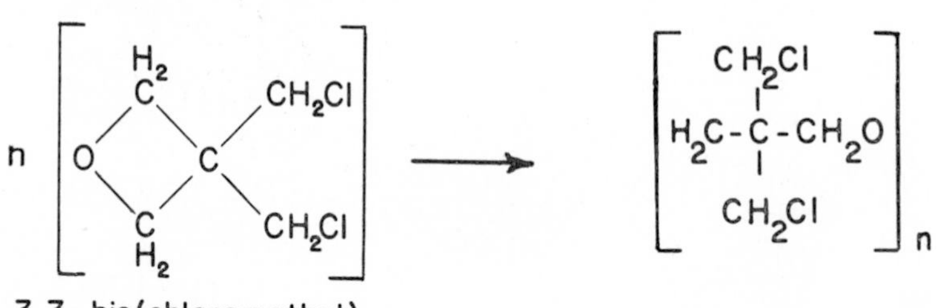

3,3—bis(chloromethyl) oxetane

"Penton"

Fig. 13.4—Polymerization of chlorinated polyether.

method, by flame spraying, by spraying as a solution in an organic solvent or a suspension in water. Equipment such as pumps and meters may be made either of metal coated with the resin or solid molded resin parts.

Trademark name: Penton

PHENOXIES

Production

Phenoxy resins are high molecular weight (about 30,000) polyhydroxyethers made by reacting bisphenol A and epichlorohydrin (see Figure 13.5). They differ from epoxy resins since they contain no epoxy groups and are higher in molecular weight. They are thermoplastic resins with a linear structure.

Properties and Uses

Phenoxy has high rigidity and impact strength. It has very good creep resistance, high elongation, low moisture absorption, very low gas transmission, and it is self-extinguishing. It can be injection or blow-molded, extruded, and applied as a coating from solution or in a fluidized bed. Although normally used as a thermoplastic it can be cross-linked with iosocynates, anhydrides, triazines, and melamines.

Phenoxy resins are excellent adhesives for wood and several metals. They are used in various protective coatings for both wood and metals. In the molded form they are used in electronic parts, sporting goods, and appliance housings. They are extruded into pipe which is particularly suitable for handling gas and crude oil.

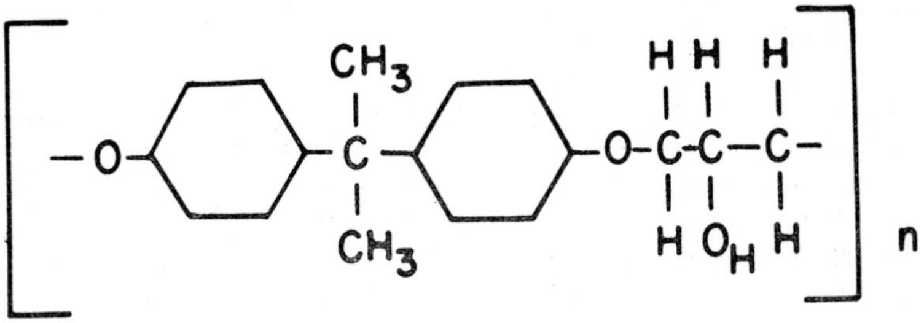

Fig. 13.5—Phenoxy formula.

CHAPTER 14

Rubbers

The rubbers, or elastomers, are differentiated from the other plastic materials by their property of reversible extensibility. The "vulcanized" products, except for hard rubber, may be deformed readily but return rapidly to their original dimensions when the deforming load is removed. These elastomers have a considerable range of properties (see Table 14.1), and the dividing line between them and other plastics is not sharp.

The principal elastomers may be considered as polymers or copolymers of 1,3 butadiene ($CH_2=CH\ CH=CH_2$) or one of its derivatives. Isoprene ($CH_2=C\ CH\ CH_3=CH_2$), the natural rubber monomer, is the 2-methyl derivative. Chloroprene ($CH_2=CHCCl=CH_2$), the neoprene monomer, is the 2-chloro derivative.

Another group of elastomers is composed of ethylene derivatives or copolymers. Still other elastomers are polyurethanes, polysulfides, and polyacrylates.

Prior to World War II, except for a relatively small amount of neoprene used in special applications, natural rubber was the only elastomer used in appreciable amounts in the United States. During the war, with the supply of natural rubber from Asia cut off, the manufacture of styrene-butadiene copolymer rubber was started in government plants in this country as an emergency measure. The production of this and other types of rubber has increased rapidly since the war.

Table 14.1: Comparative Properties of Rubbers

Property	Natural	SBR	Butyl	Polysulfide*	Nitrile	Neoprene	Chlorosulfonated polyethylene†	Fluorocarbon‡	Urethane§	Ethylene-propylene‖	Silicone
Tensile Strength (1000 psi)	3.0	2.0	2.0	...	2.0	3.0	3.0	2.0	4.0	3.0	1.5
Specific Gravity	0.9	0.9	0.9	1.3	1.0	1.2	1.2	1.9	1.1	0.9	...
Vulcanizing Properties	E	E	G	F	E	E	E	G	E	E	...
Adhesion (to metals & fabrics)	E	G–E	G	F–P	G	E	G–E	G–E	E	G	...
Tear Resistance	G	F	G	P	F	G	G	F	E+	G	P
Abrasion Resistance	E	G–E	G	P	G	E	E	G	E+	E	P
Gas Permeability	F	F	VL	L	F	L	L–	VL	F	F	F
Dielectric Strength	E	E	E	F	P	G	E	G	E	E	G
Chemical Resistance:											
Acids	F–G	F–G	E	F	G	G–E	G–E	E	P	E	F–E
Petroleum products	P	P	P	E	E	G	G	E	E	P	F
Fat and oil solvents	P–G	P–G	E	E	E	G	G	E	E	G	F
Aliphatic	P	P	P	E	E	F	F	E	E	P	P
Aromatic	P	P	P	G	G	P	P	E	F–G	P	P
Oxygenated	G	G	G	G	P	P	P	P	P	G	F
Lacquer	P	P	P	G	F	P	P	P	P	P	P
Resistance to:											
Oxidation	G	G	E	G	G	E	E+	E+	E	E	E
Sunlight	P	P	VG	G	P	VG	E+	VG	G	E+	E
Heat	G	E	E	P	E	E	E	E	G	E	E
Cold	E	E	G	F	G	G	G	G	E	E	E

P=poor; F=fair; G=good; E=excellent; L=low; V=very.
* Thiokol
† Hypalon
‡ Viton
§ Adiprene
‖ Nordel

At the present time (1966) only about 25 per cent of the rubber consumed in the United States and 45 per cent consumed in the world is natural rubber. About 70 per cent of the synthetic rubber used in this country is the styrene-butadiene copolymer made largely by the emulsion polymerization method. The use of the new stereospecific catalysts in solution polymerization have made possible better control of the rubber structure and the use of new raw materials. The trend appears to be away from the styrene-butadiene product to some of those produced by the newer methods.

ISOPRENE RUBBERS

Natural Rubber

Sources. Natural rubber is the cis 1,4 polymer of isoprene. Although there are about 490 species of rubber-producing plants, our natural rubber today comes almost entirely from the cultivated Hevea rubber trees (see Figure 14.1). Most of this rubber is produced in southeastern Asia with small amounts in South America and Africa. The shrub *guayule* which grows on dry land in southwestern United States and in Mexico has been investigated as a rubber source, but never commercialized. The various species of milkweeds contain some rubber.

The rubber from the Hevea tree occurs as an emulsion or dispersion, known as latex. The rubber, which constitutes about 35 per cent of the latex, occurs as particles 0.1 to 2 microns in diameter which are kept dispersed in the water by about 2 per cent protein material. A downward spiral cut is made in the bark of the tree (see Figure 14.2). The latex flows out and down the cut into a cup from which it is collected. The latex may be concentrated by centrifuging or evaporating and is then shipped to the factory for certain applications for which it is preferred to the dry rubber. Most of the latex is treated with acetic or formic acid to precipitate the rubber. The separated rubber was formerly formed into sheets and dried by hot air or smoke (see Figure 14.3). A more modern method uses extruders to squeeze out the water. In one method the coagulated rubber passes through three or four pairs of creping rolls which cut it into crumb form. The crumbs are baled for shipment.

Processing. Raw rubber is milled between heavy rolls or in a Banbury mixer (see Figure 2.5) to secure uniformity and to

Fig. 14.1—Young rubber trees. (Courtesy Natural Rubber Bureau)

soften it so that it can be mixed with additives and processed in operations such as calendering and extruding. Certain plasticizers, both chemical and mechanical in their operation, are added to aid in the softening. Among the chemical plasticizers are certain petroleum sulfonates, aromatic mercaptans, and pentachlorothiophenol, as well as vulcanization accelerators which may also act as plasticizers. "Physical" plasticizers which penetrate between the polymer chains, spreading them apart mechanically, include petroleum oils and waxes, pine tars, coal

Fig. 14.2—Tapping a rubber tree. (Courtesy Natural Rubber Bureau)

Fig. 14.3—Smoked and dried rubber being wheeled from the smoke-house. (Courtesy Natural Rubber Bureau)

tar oils and pitch, vegetable oils, and fatty acids, as well as synthetic organic plasticizers used in other types of plastics.

The raw rubber chains must be cross-linked sufficiently to develop the desired reversible extensibility. Sulfur is commonly used for this cross-linking or "vulcanizing." In ordinary soft rubber about 4 per cent or less sulfur is used, thus linking from 5 to 10 per cent of the double bonds in the chain. These cross-links increase the tensile strength from around 140 psi to as much as 4200 psi. Vulcanized rubber is attacked by solvents to a lesser degree than the raw rubber since the cross-links prevent the chain from separating more than a definite amount. If about 45 per cent sulfur is used in vulcanizing, hard rubber, a thermally set rigid product without plasticity, is produced.

The sulfur is mixed with the softened rubber and the other additives in a Banbury mixer. Since the time required for vulcanizing rubber using sulfur alone is several hours at 280° F, accelerators are commonly added. This allows a reduction in vulcanization time to a few minutes. Representative accelerators are mercaptobenzothiazole, benzothiazyldisulfide, diphenylguanidine, and tetramethylthiuramdisulfide. To reduce the susceptibility of the finished product to attack by oxygen at the remaining double bonds, antioxidants are added. These antioxidants are of three types: (1) secondary amines, (2) phenolic compounds, and (3) phosphites.

Fillers such as clays, silica, calcium carbonate, calcium silicate, and carbon black are also added. While fillers tend to reduce the cost of the rubber product, they may also add or increase desirable properties. The most important filler is the carbon black, added in amounts up to 50 parts for each 100 parts of rubber. The beneficial effect of the carbon black depends upon both the quantity and the quality. The latter is associated with particle size, structure, and chemical properties.

Various pigments may be used to color the rubber. Zinc oxide is not only a good white colorant but also acts as a reinforcing filler. Lithopone and titanium dioxide are also used as white pigments. Inorganic colored pigments include iron oxides and ultramarine blue. Organic pigments need to be carefully selected to avoid those which are sensitive to molding conditions or to chemicals included in the mix.

Application Methods. Rubber is formed by compression and transfer molding and by extrusion. The usual temperature range of 260° to 340° F for vulcanization makes the use of steam for heating practical. As the vulcanization, or curing, time increases, the properties change but not necessarily at the same rate. Thus there is no one time of cure at which all the characteristics of rubber are at their best values. The curing time of each formulation is set to give the product best suited for a given end use.

Automobile inner tubes are extruded and then cured, after inflation, in a steel mold. Other extruded pieces may be cured in batches in autoclaves. Electrical wire is coated by extrusion and vulcanized continuously as it moves through a steam-heated tube, perhaps 200 feet long. Similar continuous curing processes for other extruded products are being developed.

In the production of automobile tires, parallel horizontal cords are first coated on the bottom and the top with a sheet of unvulcanized rubber mix. The sheets with the cords between are formed into single continuous plies between rolls. The carcass of the tire is then built up from the plies over a form. The extruded tread is placed on top and extruded strips on the sides (see Figure 14.4). The tire is cured in a hot steel mold under pressure supplied by an inflated rubber bag resembling a heavy inner tube (see Figure 14.5).

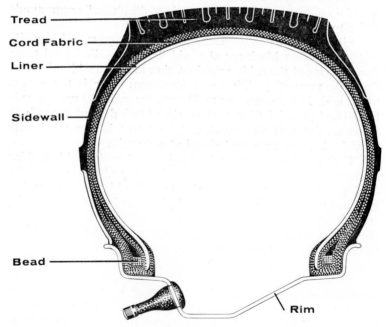

Tread

Cord Fabric

Liner

Sidewall

Bead

Rim

Fig. 14.4—Parts of a tire. (Courtesy Firestone Tire and Rubber Company)

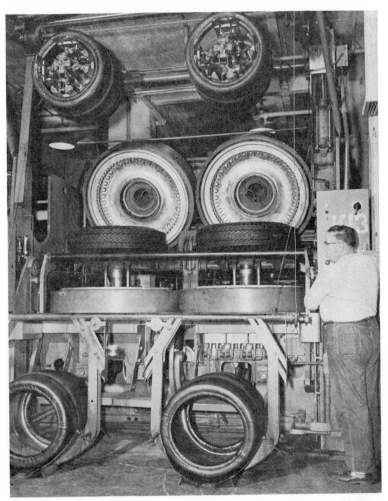

Fig. 14.5—Vulcanizing tires. (Courtesy Firestone Tire and Rubber Company)

Another type of tire known as the "radial-ply" tire has the cords running at a 90° angle to the direction of travel in contrast to 36° for the conventional tire. The carcass of the tire is readily made with natural rubber, but not all synthetic rubbers can be satisfactorily used.

Some rubber products are made directly from the latex. Concentrated latex containing 62 to 68 per cent total solids is usually used. The various additives such as are used in dry rubber compounding are emulsified before blending with the latex. Carbon black is seldom used since it serves no useful purpose in latex products. The compounded latices are formed into shape by such techniques as dipping, spreading, spraying, and molding. The films thus formed are coagulated by dilute acids or solutions of divalent salts. Sometimes the water is removed by evaporation. Vulcanization is carried out with steam or hot air. *Sponge* rubber is made from dry rubber by the use of chemical blowing agents. *Foam* rubber is made by incorporating gas into latex.

Properties and Uses. Natural soft rubber is used in applications where both its mechanical properties, such as flexibility and extensibility, and its resistance to chemicals make it particularly suitable. These include such well-known objects as automobile tires and tubes; water, steam, and air hose; "rubbers" and overshoes; raincoats; floor tile and mats; gaskets; insulating wire covering; tubing; protective gloves; toys; hot water bottles; and erasers. Industrially it has many uses such as protective lining for tanks, pumps, and other equipment; belting; and various types of seals and closures. Hard rubber is used in such items as combs, electrical insulating parts, and chemically resistant pipes and equipment.

Reclaimed Rubber. Discarded rubber items, especially tires and tubes, are treated to recover the *reclaimed rubber.* In the common process the rubber items are ground between corrugated rolls revolving at unequal speeds. The ground rubber material is digested 5 to 24 hours in an aqueous solution of sodium hydroxide at from 370° to 405° F. This hydrolyzes the fiber and softens the rubber. After being washed and dried, the rubber is milled and blended with plasticizer, clay, and carbon black. After further milling and straining to remove foreign matter, the rubber is formed into thin sheets which are wound into rolls.

Reclaimed rubber is blended with new rubber in many rubber products. Among the advantages given for it are low cost, short mixing time, low power consumption, low heat development in processing, high rate of cure, and good aging.

Balata and Gutta-percha. These rubbers are secured from tropical trees in much the same manner as Hevea rubber, but have the *trans* instead of *cis* structure (see Figure 14.6). They can be

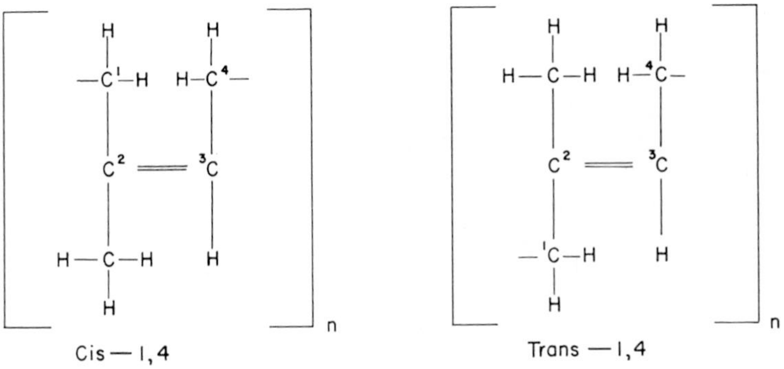

Fig. 14.6—Cis and trans rubber formulas.

blended with Hevea rubber and vulcanized. They have been used as a covering for golf balls, insulation for submarine cables, and in sheeting and tubing.

Synthetic Polyisoprene

Cis 1,4 Polyisoprene. This synthetic "natural" rubber is made by polymerizing isoprene, using stereospecific catalysts such as a titanium or a Ziegler type. The monomer is commonly separated by extractive distillation from the other thermal cracking products of gas oils and naphtha. The isoprene, dissolved in n-pentane, is polymerized in a stainless steel jacketed reactor at approximately 120° F. To insure purity both the isoprene and solvent are redistilled and dried before use.

The reaction is exothermic and requires the removal of considerable heat. The mixture of polymer and solvent is very viscous, causing mixing and handling problems. At the end of the reaction the batch is pumped into a holding tank where catalyst deactivator and antioxidant are added. The solvent is then vaporized off and condensed for reuse.

The product is very similar to Hevea rubber which has undergone preliminary "breakdown" in processing. This makes it easier to process. It can be used to replace natural rubber for most uses and is becoming competitive in price. Napthenic oils are commonly added to the extent of 25 parts per 100 parts of the rubber. The oils act as plasticizers and "extenders" as well as to lower the cost.

Butyl Rubber

Production. Butyl rubber is a copolymer of isoprene and isobutylene [$H_2C{=}C(CH_3)CH_3$]. Isobutylene is separated from the C_4
petroleum fraction by absorption in sulfuric acid, from which it
is stripped by steam.

Butyl rubber is an example of a polymer designed to have
a definite amount of cross-linkage. The isobutylene has only one
double bond and therefore can form only a chain compound
without any cross-linkage. By inserting the proper amount of
isoprene and cross-linking at all double bonds, a saturated
polymer with desired plasticity and strength can be produced.
The lack of available double bonds in the vulcanized rubber
makes it more resistive to oxygen and other chemicals than
natural rubber.

The production of butyl rubber is somewhat complicated
by the low temperature at which it must be carried out. A solution of 25 per cent isobutylene, 2 to 3 per cent isoprene, and
aluminum chloride in methyl chloride is reacted at —140° F.
After flashing off the unreacted monomers and the methyl
chloride by heating, the polymer crumb is dried and formed into
sheets.

Processing. Butyl rubber is processed by the same general methods used with natural rubber. Because of the low unsaturation,
ultra-accelerators such as derivatives of dithiocarbamic acid are
desirable in vulcanization. One to two parts sulfur are used.
Heating for 2 to 4 hours at about 320° F between milling cycles
is said to improve the elasticity and resilience. As much as 50
parts oil and 75 parts carbon black can be satisfactorily added to
butyl rubber to form the "master batch."

Properties and Uses. Butyl rubber is made with unsaturation
varying from 1 mole per cent to 2.5 mole per cent isoprene.
Ozone and other chemical resistance decreases with increase in
unsaturation from excellent to good. Increase in unsaturation
improves heat resistance and speeds up curing.

Butyl rubber, because of its low gas permeability, has been
a preferred material for inner tubes and as an inner air barrier
in tubeless tires. It is also used in air cushions, pneumatic
springs, accumulator bags, and air bellows. Because of its excellent electrical properties it is used as insulation on power
cables and other electrical products. Because of its resistance to
weathering it is used in convertible car tops, wading pools, and

raincoats. It is used in mechanical rubber goods such as steam and acid hose, tank linings, belts, and many parts for cars. It is being used in increasing amounts in tires where it is said to have some advantages over natural rubber and butadiene-styrene rubber.

Chlorinated butyl rubber is produced by selective chlorination of the isobutylene-isoprene copolymer. The product which contains about 1.2 per cent by weight of chlorine differs little in unsaturation from the original polymer. Unlike the unchlorinated product, this modified butyl rubber can be blended with other rubbers. It can be cross-linked by other reagents than sulfur, thus varying its chemical resistance to heat, oil, and ozone.

Natural Rubber Derivatives

Rubber Chloride. Chlorine is bubbled into a solution of natural rubber in a solvent such as carbon tetrachloride at 175° to 230° F. The granular product is separated by running into hot water, distilling off the solvent, and centrifuging. It is used in special-purpose paints and as a bonding agent for metals. A similar product can be made from synthetic polyisoprene and butadiene-styrene rubbers.

Rubber Hydrochloride. This product can be made by bubbling hydrogen chloride into a solution of rubber or by treating rubber with liquid hydrogen chloride at —26° to —45° F. This is used largely as a packaging film.

Isomerized Rubber. When rubber is heated with acids such as sulfuric, benzenesulfonic, or p-toluene sulfonic, a rearrangement occurs in the molecule, producing "thermoprene," used in a cement for bonding rubber to metals and in chemically resistant paints. A similar compound is produced by reacting rubber with certain metallic chlorides, such as stannic or titanium chlorides. It is used as a chemically resistant tank lining, in paints, and as a thermoplastic molding compound.

BUTADIENE RUBBERS

Butadiene-Styrene Copolymer

Production. While butadiene can be polymerized to form a straight chain polymer, the original products were difficult to process. The first commercial butadiene rubber was the copoly-mer with styrene or vinyl benzene ($C_6H_5CH=CH_2$). The branch-ing added to the molecule by the styrene apparently made it easier to process, although still more difficult than natural rub-ber. The common copolymer contains approximately one mole of styrene to six of butadiene. The styrene is produced by de-hydrogenating ethyl benzene which is produced by reacting ben-zene and ethylene with aluminum chloride as a catalyst.

Polymerization was originally carried out in an emulsion at 120° F for 12 to 15 hours in large glass-lined reactors. It was later found that the "cold rubber" produced at about 40° F had a higher molecular weight and had substantially improved abra-sion resistance when used as tread stock. Polymerization is in-itiated by free radicals produced by a redox pair. Usually fer-rous pyrophosphate, soluble in water, is the reducing agent and p-menthane hydroperoxide soluble in the monomer is the oxi-dizing agent. A more recent development is solution polymeriza-tion using a stereo specific catalyst. This is said to produce a bet-ter product. A variation of the cold rubber is the *oil-master-batched* rubber. From 25 to 50 parts of a petroleum oil in emul-sion form are added to each 100 parts of polymer in the reactor before coagulating it. The higher molecular weight of the cold rubber allows it to be successfully extended with the oil. Carbon black is sometimes also added as a water dispersion to the latex.

Processing. The compounding and processing of butadiene-styrene rubbers are similar to those operations using natural rub-ber. The butadiene-styrene rubbers are less unsaturated than the natural rubber and are slower in curing. Milling requires somewhat more power. The unvulcanized butadiene-styrene rub-ber is not tacky like natural rubber. In tire manufacture this makes it necessary to use natural rubber cement on the surfaces to be joined.

Uses. Butadiene-styrene rubber is used for practically all uses of natural rubber. The original rubber was not as satisfactory for tire treads, particularly for large truck tires, as the natural rub-ber. The newer cold rubber with a newer carbon black as rein-forcing filler appears to be at least as good if not better than natural rubber for tires. It oxidizes less readily and is less liable to cracking than the natural rubber. Butadiene-styrene hard rub-ber is very similar to natural hard rubber.

Nitrile Rubber

Production. Nitrile rubber is the copolymer of butadiene and acrylonitrile, or vinyl cyanide ($H_2C{=}CHCN$). The acrylonitrile can be produced by reacting hydrogen cyanide and acetylene, using an aqueous solution of ammonium chloride and cuprous chloride as a catalyst.

The copolymerization of butadiene and acrylonitrile is carried out in an emulsion at from 40° to 75° F in much the same manner as for a butadiene-styrene rubber. The copolymer can be modified by the addition of methacrylonitrile, styrene, vinylidene chloride, methyl methacrylate, and the acrylic acids.

Processing. Processing is similar to that for natural or butadiene-styrene rubbers. The nitrile rubber is less plastic than natural rubber and develops more heat in milling. The choice of fillers and pigments is more important in developing satisfactory properties than in natural rubber.

Properties and Uses. Desirable characteristics of nitrile rubbers include excellent oil and heat resistance. Resistance to paraffinic oils is greater than to aromatic oils. While the resistance to oils increases with acrylonitrile content, the increase is small through the usual range of 18 to 45 per cent. Increasing the acrylonitrile to 50 or 60 per cent produces a leathery product with high resistance to even aromatic oils.

Since nitrile rubbers are higher in cost than natural rubber, they are used mainly in special applications such as bullet-sealing tanks and fuel hose where oil resistance is needed. They are used in blends with polyvinyl chloride which have greater sunlight and ozone resistance than the rubber alone. These blends are used in upholstery fabric, insulation, and leather substitute. Phenolic resins are blended with nitrile rubber to produce a stiffer, stronger product. High percentages of the resin produce thermosetting adhesives used for bonding metals, plastics, and glass. Nitrile rubber also blends with natural rubber, high-styrene copolymers, and polysulfide rubbers. In general, it tends to add toughness.

When nitrile rubber is vulcanized with sulfur corresponding to one sulfur atom per double bond, a hard rubber is produced. The strength of this hard rubber is less affected at high temperatures than natural hard rubber. However, it has less impact strength and higher electric power loss than natural hard rubber.

Modified Nitrile Rubber. A new modified nitrile rubber has acrylic-type acids added to the monomers. This introduces carboxyl groups along the rubber chain with about one for each 100 to 200 carbon atoms. This modified rubber, in comparison with a nitrile rubber of equal oil resistance, has better tensile strength, hardness, low-temperature brittleness, and ozone resistance. Blends with vinyl resin produce a product with excellent strength and low-temperature properties.

Polybutadiene Rubbers

Cis Polybutadiene. Butadiene is polymerized into a chain polymer, using a stereospecific cobalt catalyst of the Ziegler type. Polybutadiene can be processed by the conventional rubber methods and can be blended effectively with natural and styrene-butadiene rubbers. Oils, fatty acids, and resin are used as plasticizers. Carbon black is used as a filler and sulfur as the vulcanizing agent.

The addition of polybutadiene to natural rubber or styrene-butadiene is said to give a longer-wearing tire tread. It is also being used alone as tread stock and is recommended for conveyor belt covers and other uses where high abrasion and flex-cracking resistance are desirable.

Trans Polybutadiene. Butadiene can be polymerized using stereospecific catalyst to give the trans structure. This is being developed to substitute for the natural trans polyisoprene rubbers, balata, and gutta-percha.

Neoprene Rubber

Production. Neoprene rubber is composed largely if not entirely of the trans polymer of chloroprene, or 2-chloro-1,3-butadiene ($CH_2{=}CCHCl{=}CH_2$). Chloroprene is made by treating monovinyl acetylene, produced by polymerizing acetylene, with hydrochloric acid. Common molecular weight range is 100,000 to 500,000.

The emulsified chloroprene is polymerized at 104° F with

potassium persulfate as a catalyst. Tetraethylthiuram disulfide is used as a short-stop to control polymerization. Acetic acid is added and the emulsion coagulated by freezing.

Processing. Neoprene is processed by natural rubber procedures. Metallic oxides, such as a combination of magnesium and zinc oxides, are superior to sulfur as vulcanizing agents. Fillers used are mainly those used in natural rubber. Petroleum oils and polyester plasticizers may be added.

Modified Varieties. A copolymer with a small amount of styrene is less crystalline. Another with isoprene is suitable for low temperatures. Copolymerizing with acrylonitrile produces a rubber with good processibility and oil resistance.

Properties and Uses. Neoprene has good strength and resistance to weathering. Its high chlorine content imparts a high resistance to heat and flame. This, in addition to excellent resistance to oil and chemicals, makes neoprene suitable for many industrial applications such as gasoline hose and special conveyor belts. The different varieties combined with various fillers and other modifying agents make possible special properties for individual applications. Neoprene latices are used in production of gloves, balloons, industrial parts, and foam items. They are also used in adhesives.

ETHYLENE RUBBERS

Ethylene-Propylene Rubbers

Production. Mixtures of propylene with 20 to 80 per cent ethylene can be copolymerized in a solvent at between 50 and 150 psi with a catalyst such as dialkylaluminum combined with vanadium oxytrichloride to form an elastomer (EPR). This can be vulcanized with a peroxide such as dicumyl peroxide. The terpolymer (EPT), which can be vulcanized with sulfur, is made by adding about 3 per cent dicyclopentadiene to the ethylene and propylene and polymerizing in hexane as a solvent. A typical formulation might be 50 parts each of ethylene and propylene with 3 per cent dicyclopentadiene. Experimental work has shown that the terpolymer can also be cured with phenol-formaldehyde resins with a heavy metal catalyst. Another method of producing a readily vulcanizable elastomer is to chlorinate the copolymer.

Processing. The copolymer and terpolymer are readily processed in conventional rubber machinery and are compatible with oils, carbon black, and mineral fillers.

Properties and Uses. These rubbers are highly resistant to ozone degradation, have good electrical properties, excellent abrasion resistance, and low density.

The terpolymer is being used in increasing amounts in tires, wire insulation, and hose. It is a good general purpose rubber.

Trademark name: Nordel

Chlorosulfonated Polyethylene Rubber

Production. This rubber is made by substituting chlorine and sulfonyl groups into polyethylene with a molecular weight of about 20,000. Approximately one chlorine atom to each 7 carbon atoms and one sulfonylchloride group for each 90 carbon atoms are added.

The saturated rubber is cured with litharge (PbO) or magnesia (MgO) and an accelerator such as mercaptobenzothiozole.

Uses. This rubber can be extruded, molded, or formed to sheets by calendering. It is also available as a coating (paint) material. It has excellent temperature resistance and resistance to many chemicals, including oxidizing acids.

It is used for lining of chemical tanks, for hose for handling acids, for conveyor belts, and in a variety of molded products where resistance to heat and chemicals is important. It is used as a coating over other rubbers to protect against ozone. Coated fabrics are used in a variety of products such as industrial pump diaphragms, camera bellows, and automobile topping.

Trademark name: Hypalon

OTHER RUBBERS

Polysulfide Rubbers

Production. Polysulfide rubbers are condensation products of an organic dihalide with a polysulfide. Typical combinations are ethylene dichloride and sodium polysulfide, and dichlorethyl formal and sodium sulfide. Properties of the rubber may be modified by using two or more chlorides together and other sulfides may be substituted. Bromides may also be substituted for the chlorides. The constituents are reacted in water solutions in stainless steel agitated reactors. A latex results from which the rubber may be coagulated by the addition of acetic or sulfuric acid.

Properties and Uses. Polysulfide rubbers vary in properties but in general have lower tensile strength and abrasion resistance than natural rubber, but excellent weathering, oxygen, and oil resistance. Most solvents do not dissolve them, although some cause swelling.

The polysulfides are used for o-rings, gaskets, printing rolls, and as lining in bullet-proof airplane gasoline tanks and in hose. They are used as protective coatings, applying either in the latex form or by flame spraying.

It is possible to break down the rubber to a lower molecular weight liquid product by milling with plasticizers such as benzothiazyl disulfide. The liquid polymers may be used for castings, impregnation coatings, and adhesives. Curing is by lead oxide or organic peroxides.

Acrylic Rubbers

Production. Acrylic rubbers are copolymers of ethyl, methyl, or butyl acrylate. A typical rubber is about 95 per cent ethyl acrylate and 5 per cent chloroethylvinyl ether. The acrylic rubbers are produced by emulsion polymerization. Another possibility is a copolymer with ethylene.

Processing. Acrylic rubbers are vulcanized or cured with amines such as triethylenetetramine, and tetraethylenepentamine.

Carbon black is used as a reinforcing filler in amounts from

35 to 60 parts per 100 parts of the polymer. White silica types of pigments can also be used in similar amounts. Softeners or plasticizers are used only to a limited extent. "Tempering" of vulcanized articles at around 300° F for 24 hours improves the properties of finished articles.

Properties and Uses. Acrylic rubbers have excellent resistance to high temperatures and to oils. They can be used at temperatures from —10° to 400° F. They have very good resistance to oxygen, ozone, and sunlight. They are decomposed by alkaline solutions and have poor resistance to acids. The tensile strength of the acrylic rubbers is from 500 to 2400 psi and elongation 100 to 400 per cent.

The acrylic rubbers are used mainly in applications where their excellent resistance to heat and to oils, especially sulfur-containing oils, is needed. These applications include oil hose, automobile gaskets, o-rings, belting, tank linings, and cements.

Acrylic rubbers are also available in latex form. Uses include sizes and binders for textiles, paper, and leather, and adhesives where the bond is exposed to oil.

Trademark names: Hycar PA, Lactoprene EV

Fluoroelastomers

Properties. These rubbers are copolymers of vinylidene fluoride and hexafluoropropylene. Their outstanding characteristics are their excellent resistance to most oils, chemicals, and solvents up to 400° F. They have good strength and adherence to fabrics and metals. Abrasion, flame, ozone, sunlight, and weathering resistance are all very good. They are readily processed on ordinary rubber equipment.

Uses. Because of high cost the fluoroelastomers are not used in applications where natural rubber or one of the common synthetic rubbers are suitable. Uses include precision seals, o-rings, diaphragms, tubing, hose, linings, and coated fabric. A specific use is as o-rings and valve seats in refining valves handling toluene, xylene, alkylate gas, and propane at temperatures from 300° F to 400° F. Elastomer-coated asbestos is used for fire seals in the wings of a new Air Force plane.

Trademark name: Viton

Urethane Rubbers

Production. These elastomers result from the reaction of iso-cyanates with active hydrogens in polyols, such as polyesters, poly-ethers, glycols, or castor oil. They contain the urethane linkage ($RNHCO_2R'$). Commonly used isocyanates are toluene diiso-cyanate and hexamethylene diisocyanate.

The first urethane rubbers were produced in Germany by reacting linear polyesters and naphthalene diisocyanates, and were followed by a British elastomer based on polyesteramides and diisocyanates. Several similar American products are avail-able. The original polyurethane rubbers were unstable. To avoid this, one type of rubber is made with less than the equiva-lent amount of diisocyanate, and vulcanized by heating with ad-ditional diisocyanate. In another product an unsaturated olefin is incorporated. This rubber can be vulcanized by sulfur or by dicumyl peroxide. Another product (Estane) is a thermoplastic polyester-urethane elastomer used without vulcanizing. Still another type is a liquid urethane elastomer produced by reacting a linear hydroxyl-terminated polyester or polyether glycol. This prepolymer is isocyanate-terminated and the chain can be ex-tended and cross-linked by a polyamine or polyhydroxy com-pound.

Properties and Uses. Most of the urethane rubbers can be rein-forced with carbon black similarly to other rubbers. They have high strength and load-bearing capacity, very good resistance to abrasion, tear, and oils.

Uses include solid tires for industrial trucks, vibration mountings, belts, shoe soles and heels, seals, gaskets, and gears. Large amounts are used in foams. A more recent use is in elastic fibers.

Trademark names: Adiprene, Chemigum SL, Durethene, Dyalon, Genthane, Texin, Vulcollan

Silicone Rubbers

Production. Silicones (see Chapter 12) are compounds containing silicon, oxygen, and one or more organic groups. The silicones with elastomeric properties are of four types. The general pur-pose rubber is the polydimethylsiloxane made up of alternating silicon and oxygen molecules with two methyl groups attached to each silicon. The second type has a small percentage of vinyl groups substituted for part of the methyl groups. This insertion of an unsaturated group into the polymer simplifies vulcanization and increases the applications of the rubber. The third type has phenol groups substituted for a small part of the methyl groups. This lowers the brittle point, improving the low temperature properties. The fourth type, which is relatively new, is a fluorosilicone, polytrifluoropropylmethyl siloxane. It has physical properties similar to the other silicone rubbers but has superior solvent resistance.

Processing. Processing of the silicone rubbers is very similar to processing natural rubber except that no accelerator, softener or plasticizer, or antioxidant is required. The vulcanizing or curing agent is benzoyl peroxide which breaks down above 185° F to produce free radicals. These remove a hydrogen from a methyl group, producing reactive —CH$_2$ units which will cross-link with other like units. The rubber sets up in the mold after 5 or 10 minutes at 260° F. It is then necessary to heat it for several hours at 480° F to complete the cure and drive off any volatile impurities. Various manufactured and natural silica products are used as fillers. Carbon black may be used with the unsaturated methylvinyl rubber. The rubber can be compression molded, extruded, or formed into sheets.

Uses. Silicone rubbers are used mainly in applications at extremely high or low temperatures. Typical uses are wire and cable insulations, gaskets and seals for electric and electronic equipment, and seals in outdoor floodlights. A large amount of silicone rubber is used as seals, gaskets, and o-rings in aircraft, particularly military craft. Silicone rubber vibration dampeners and shock mounts for electronic equipment are used where subjected to temperature extremes. They are effective at a range of —50° to 300° F.

Liquid RTV (room temperature vulcanizing) rubber is supplied as a liquid to which a catalyst is added before using. This vulcanizes at room temperature in from 10 minutes to 24 hours to form a rubbery silicone solid, with good physical and dielectric properties. It is used as a caulking and sealing material, a potting compound, and an encapsulating material. It is especially suitable for electronic equipment in airplanes and missiles. It is also used to make flexible molds in which plastic or metal parts may be cast (see Figure 14.7). The silicone mold will withstand metals or alloys melting as high as 500° F.

Silicone rubber has limitations. It is not as strong as natural or butadiene-styrene rubber nor does it have the gasoline resistance of neoprene or nitrile rubbers. The high cost of silicone rubber prevents its use where other rubber would serve as well or better.

Trademark name: Silastic

Epichlorohydrin Rubbers

These elastomers are available in two forms: the homopolymer of epichlorohydrin and the copolymer of epichlorohydrin and ethylene oxide (32 per cent). They are said to have exceptional resistance to ozone, oil, solvent, and heat. The homopolymer is better than butyl in impermeability and is self-extinguishing. The copolymer has less flame resistance but a greater service temperature range of —40° to 300° F.

Both rubbers are readily processed by usual methods, using carbon black and other fillers and reinforcers. Piperazine hexahydrate and 2-mercaptoimidazoline are recommended curing

Fig. 14.7—Removing silicone rubber mold from a casting. (Courtesy Dow Corning Corporation)

agents. Both rubbers are used in oil resistant seals and hoses and other applications requiring resistance to oil, oxygen, and high temperature.

Propylene-Isoprene Polymer

An alternating polymer of propylene and isoprene has been synthesized experimentally, using Lewis acid catalysts. It is said to show considerable promise, partly because of low cost.

CHAPTER 15

Fibers, Films, and Foams

Three general forms of plastics—fibers, films and sheets, and foams and other cellular products—have developed to the extent that it seems desirable to consider these as one group in a common chapter. It is hoped that this will make comparisons of items of each form but of different compositions simpler than if the descriptions of these items were scattered through the various chapters dealing with other end products. For the most part the polymers used in the production of the fibers, films, and foams have been described in other chapters.

FIBERS

Synthetic fibers made from various plastic polymers are used in products varying from heavy tow ropes to fine textiles. They compete with natural fibers sometimes on a cost basis, sometimes on a quality basis, and sometimes on both. Some synthetic fibers are not attacked by moths, bacteria, and fungi as are certain natural fibers. Some synthetic fabrics have superior press-retaining properties. On the other hand synthetic fibers frequently have low softening temperatures which require caution in ironing. Some people are allergic to certain synthetics.

Synthetic fibers are produced by extrusion. Coarse fibers for cordage or woven seat covers are made as monofilaments, a

159

single filament extruded as a unit. Textile fibers may be mono-filament or multifilament. Multifilament fibers are made by ex-truding more than one filament through small openings in a die and then twisting these to form a single thread. The fiber may be extruded from molten polymer, as nylon, and then cooled to harden. Another method is to extrude a solution of the poly-mer in a volatile solvent, as cellulose acetate in acetone, and evaporate the solvent. Still another method is to extrude the viscous liquid into a chemical hardening bath, as is done with viscose rayon. It is common practice to stretch the newly hard-ened filament to orient the molecules, thus increasing the strength.

Textile fibers may be either continuous or staple. Staple fibers are made by cutting into 1½- to 8-inch lengths and then spinning, much as cotton is spun. The fabric has a wool-like appearance and feel. Textile yarns are identified by the number of filaments and the "denier." The denier is the weight in grams of 9,000 meters of yarn. Tensile strength, or "tenacity," is stated as the number of grams required to break a fiber of one denier at a definite loading rate. For most use the relation between the tenacity and elongation is of more value than tenacity alone. The bursting strength is directly related to tenacity and elonga-tion.

Textile fibers are designated by descriptive, generic, and brand names. Descriptive and generic names having general ap-plication include acetate, acrylic, azlon, nylon, polyester, poly-olefin, rayon, saran, spandex, and vinyon. For data on individual fibers see Table 15.1.

Table 15.1: Properties of Fibers*

Fiber	Tenacity†	Elonga-tion at Break	Wet Strength	Ironing Temper-ature	Specific Gravity
	grams per denier	per cent	per cent of dry	°F	
Cotton	2.5–3.0	6–7	100	400	1.52
Rayon	1.5–5.0	13–15	50–75	300	1.52
Acetate	1.2–1.4	25	65	300	1.35
Acrylic	2.3–5.0	17–35	80–96	300	1.16
Polyester	4.5–7.5	11–40	100	275	1.38
Nylon	4.3–8.8	18–45	80–90	275	1.14
Saran	2.4	15–20	100	160	1.70
Protein	0.8–1.2	15–50	60	450	1.30
Polyethylene (high density)	5.0–10.0	5–50	100	220	0.96
Spandex	0.6–0.8	520–610	. . .	300	1.0

* Representative data for each class, but not necessarily all-inclusive.
† Tensile strength.

Rayon

Rayon is a regenerated cellulose fiber in which not more than 15 per cent of the hydrogens in the hydroxyl groups have been replaced in a cross-linking operation. The two types, *viscose* and *cuprammonium*, have similar properties, although the latter produces a fine yarn used in the sheerest of fabrics. Rayon yarns may be monofilament, multifilament, or staple, with a high luster or delustered by colorless pigments. Rayon is readily dyed and is resistant to mildew and moths. Textile rayon is used in a wide variety of fabrics, both alone and blended with other fibers. High tenacity rayon, made from purer cellulose with a resultant higher degree of polymerization, is used as a tire cord. Rayons can be cross-linked with urea-formaldehyde or melamine resins, glyoxal, disocyanates, and alkyl titanates to improve strength and water resistance.

Acetate

Acetate fiber is spun from a solution of cellulose acetate in acetone, hardening as the solvent evaporates. It is available in the same fiber forms as rayon. It is weaker and lower in water absorbency than rayon, but more elastic and wrinkle resistant. It is used in a wide variety of fabrics, both alone and in blends. Triacetate can be heat set and thus used in "wash and wear" garments.

Nylon

Fibers are made from both nylon 6,6 and nylon 6 by melt spinning followed by cold drawing. They have high tenacity, wet strength, extensibility, and abrasion resistance. Nylon yarns in monofilaments, multifilaments, and staples are used in hosiery and in many fabrics both alone and in blends. Nylon is used extensively in parachutes, glider ropes, transmission and conveyor belting, cordage, fish lines, and tire cord. Coarse monofilament nylon is used in toothbrushes, paintbrushes, and in tennis rackets.

Acrylic Fibers

Polyacrylonitrile alone and modified by a small amount of copolymer is made into textile fibers, mainly staple yarns. Acrylic fibers have moderate strength, low stretchability, low heat conductivity, and low softening point. Alone and blended they are used in light-weight, warm, wrinkle-resistant, and good-wearing fabrics used in both men's and women's apparel. They are also used in industrial textiles and in carpets. Similar fibers are made from a copolymer with vinyl chloride (see Figure 15.1). Another copolymer is composed of equimolecular amounts of vinylidene dinitrile and vinyl acetate.

Polyester Fibers

These are products of a polyester reaction such as that between terephthalic acid and ethylene glycol. They are next to nylon in strength, have good resistance to wrinkling, and a high degree of stretch resistance. They are used in a variety of textile products such as "wash and wear" clothes, suits, dresses, sweaters, and underwear. There are also various industrial applications. A newly developing use is in tire cord, replacing nylon.

Saran Fibers

Saran is a copolymer of vinylidene chloride and vinyl chloride or acrylonitrile. The fibers are stronger than wool or rayon with low conductivity. Their low absorbency makes them unsuitable for most clothing use. They have excellent weather resistance which makes them suitable for upholstery on outdoor furniture, automobile seat covers, and window screens.

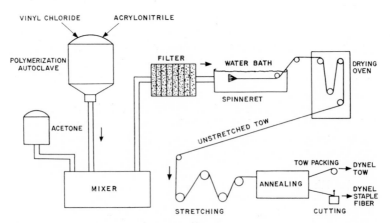

Fig. 15.1—Production of vinyl chloride and acrylonitrile copolymer fiber. (Courtesy Union Carbide Corporation)

Vinyon Fibers

Vinyon fibers are composed of polyvinyl chloride or a copolymer such as 88 per cent chloride and 12 per cent acetate. They have excellent chemical resistance and low moisture absorption, making them suitable for protective clothing, fishing lines and nets, and filter fabrics. They are also used in felts and carpets.

Polyolefin Fibers

Both polyethylene and polypropylene are extruded into fibers in the molten form, followed by cooling in a water bath (see Figure 15.2). Polypropylene apparently has better fiber-forming properties than polyethylene. While the polymers are readily colored before spinning, the fibers are difficult to dye. The polyolefins have a limited use in dress fibers and are used in protective clothing and in carpets. They are made into ropes for use on boats and in water skiing.

Elastomeric Fibers

Natural and various synthetic rubbers are extruded into filaments used with other fibers in elastic fabrics. *Spandex* elastic fibers composed of not less than 85 per cent polyurethane are superior to natural rubber in resistance to oxidation, chafing, and

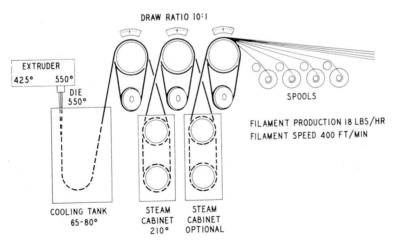

Fig. 15.2—Pilot plant production of polyethylene fibers. (Courtesy Phillips Petroleum Company)

dry cleaning damage. They have excellent resistance to abrasion, ultraviolet light, weathering, chemicals, and cosmetics and are used in foundation garments, swim wear, surgical hose, and other elastic products.

Miscellaneous Fibers

Polystyrene monofilaments are used in brushes with limited textile applications. *Polytetrafluoroethylene* fibers are used where their unusual chemical resistance and high melting point are needed.

FILMS AND SHEETS

Practically all thermoplasts are produced in film or sheet form. Films are under 10 mils thick, sheets over 10 mils.

Vinyl Films and Sheets

Plasticized polyvinyl chloride and the copolymer chloride-acetate are made into films and sheets with moderate tensile strength, good tear strength, low water absorption, and moderate chemical resistance. They can be heat sealed, and thermoformed.

Applications of film include raincoats, shower curtains, table covers, wall covering, and baby pants. Sheeting is used for upholstery, luggage, toys, seat covers, and handbags. The film and sheet can be made in a wide range of transparent, translucent, and opaque colors.

Rigid vinyl films and sheets can be cut, punched, vacuum-formed, and laminated. Uses include drafting instruments, lighting panels, signs, place mats, credit cards, recording discs, and shower doors.

Polyvinylidene Chloride Film

The plasticized film has excellent chemical resistance, low water absorption, and is heat shrinkable. It is used in packaging candies, processed and frozen meats, and other food products. It is sold for household wrapping under the name of "Saran Wrap."

Polystyrene Sheet

This can be vacuum-formed, machined, and cemented. Applications include toys, decorative panels, signs and displays, food packages, and refrigerator and freezer liners.

Expanded Polystyrene

This material is extruded into films and sheets from six to 100 mils which can be laminated and vacuum-formed. It is used for packaging where heat insulation is needed.

Polyethylene Films

This film has low density, low moisture and vapor permeability, and very good tear, impact, and tensile strength (see Figure 15.3). It is used extensively in packaging, as a moisture

Fig. 15.3—Polyethylene extruding into tube form prior to being slit into flat film. (Courtesy Phillips Petroleum Company)

barrier in building construction, in drum liners, in tarpaulins, in crop mulching, and in weather balloons (see Figure 15.4). It can be heat sealed.

Cellulose Acetate Film and Sheet

This film has outstanding clarity; is wrinkle, water, and grease resistant; has good electrical properties and stability; is readily heat formed; and can be printed. Applications include food packaging, overwraps, window envelopes, lamp shades, shades, protective covers, and vacuum-formed parts. It is used for "safety" type photographic film. The sheet is used for electrical insulation, welders' shields and safety lenses, vacuum-formed pieces, handbags, and packages. Similar films and sheets are made from cellulose triacetate and cellulose acetate-butyrate (see Figure 15.5).

Cellulose Nitrate (Celluloid)

Sheets of cellulose nitrate, plasticized usually with camphor, are sliced from a large cake. Uses include watch crystals, pipe-

Fig. 15.4—Stratospheric balloon of polyethylene film. (Courtesy General Mills, Inc.)

Fig. 15.5—Cellulose acetate-butyrate film "blisters" used for packaging. (Courtesy Chicago Molded Products Corporation)

line wrappings, photofilm, novelties, and ornaments. While tough, low in cost and water absorption, and resistant to many chemicals, it discolors and becomes brittle from sunlight and hardens with age. Its high flammability also limits its use.

Ethyl Cellulose

Unplasticized ethyl cellulose film is tough, odorless, dimensionally stable, and has excellent clarity. It is used for electrical tape, packaging, lamp shades, advertising novelties, and windows in industrial equipment.

Cellophane

Cellophane is regenerated cellulose (viscose). Sheets and films are tough and transparent and are widely used in packaging and as a base for pressure-sensitive tape. It is available in a wide variety of colors.

Polyacrylic Sheets

These sheets, made by either extrusion or casting, are fabricated by vacuum or similar thermoforming methods to form outdoor signs, displays, and airplane canopies and windows.

Polyester Film

This film is clear and tough with good dimensional stability, strength, electrical properties, and chemical resistance, except to alkalies. Applications include insulation, conveyor belts, metallic yarn, magnetic recording tape, gas barrier, and general packaging.

Trifluorochloroethylene Film and Sheeting

The film is tough, nonporous, transparent, and extremely resistant to chemicals. Uses include tank and pipe lining, gaskets, and diaphragms.

Rubber Films and Sheets

Natural and synthetic rubbers can be formed in films and sheets by calendering, extrusion, or directly from latex. Uses include a wide variety of applications such as tapes, equipment linings, gaskets, gloves, toy balloons, laboratory products, and packaging materials.

Nylon

Extruded nylon sheets and films are used for cable insulation, tapes, packaging, bearing surfaces, washers, shims, and gaskets.

FOAMS AND OTHER CELLULAR PRODUCTS

As has been pointed out (Chapter 3) expanded plastic products are made in two types of cell structure: open and closed. They are produced by (1) using blowing agents, (2) using a cross-linking agent which produces a gas, (3) by vaporizing a volatile solvent, (4) mechanically mixing in a gas, and (5) by adding expanded material to a binder.

Sponge and Foam Rubber

Sponge rubber is produced from dry natural or synthetic rubber, using a blowing agent such as sodium bicarbonate and a fatty acid. If an *open-cell* structure is desired the compounding and curing are regulated to produce the gas before vulcanizing, thus rupturing the cell walls and producing an interconnecting structure. For a *closed-cell* structure the rubber is partially cured before gas expansion. Vulcanizing is in suitable frames or molds in either batch or continuous operations.

Open-cell types of sponge rubber are used in gasketing, sealing, heat insulating, cushioning, and shock absorption. The *closed-cell* types have similar uses in addition to those requiring buoyancy in water. The hard rubber product is used in heat-insulating, sandwich construction, and floats.

Foam rubber is produced by adding gas to rubber latex and vulcanizing. It is used largely in pillows, furniture cushions, mattresses (see Figure 15.6), automotive pads, and gaskets. Toys, such as dolls, are also produced.

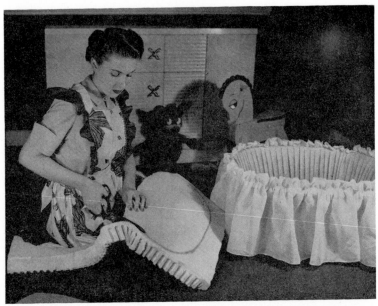

Fig. 15.6—Latex foam mattress. (Courtesy Natural Rubber Bureau)

Urethane Foams

Urethane foams may be produced by vaporizng a blowing agent or by the carbon dioxide resulting from the reaction of water and isocyanate groups in the reaction mixture. Slabs may be formed continuously on conveyors, with curing in an oven. The foams may also be formed by pouring the foaming mixture into a cavity, such as in airplane members, and allowing them to cure in place. The foams are used in thermal insulation, cushioning, and in electrical potting.

Cellular Polystyrene

Expandable polystyrene is marketed in the form of beads containing an expanding agent. When heated in a mold they expand to fill the mold. The expanded material has closed-cell structure with densities as low as one pound per cubic foot. Uses include thermal building insulation, flotation material, shipping containers, disposable drinking cups, and crash helmet liners.

Expanded polystyrene is available in blocks and sheets which are fabricated mechanically. In addition to use as a structural insulating material and in marine uses it is used in protective packaging, display platforms, and in the production of many novelty items.

Polyvinyl Chloride Foam

Three types of polyvinyl chloride cellular products are available: (1) open-cell flexible foam, (2) closed-cell flexible foam, and (3) closed-cell rigid foam. Uses are similar to those of the corresponding rubber products. Flexible vinyl foam can be laminated to fabric, producing a leather-like material used in clothing and upholstery. It can be extruded into weather-stripping and gasket material. It is injection molded into such items as sandal soles and gaskets. It can be rotationally molded to produce doll bodies and heater ducts.

Foamed Phenolics

The conventional type of phenolic foam utilizes the heat of reaction of the liquid resin polymerizing to vaporize the water and solvent. The product, which is a composite of 40 per cent open cells and 60 per cent closed cells, is used mainly as a building insulation in either preformed slabs or foamed-in-place material. Densities can be varied from $\frac{1}{3}$ pound to 80 pounds per cubic foot.

Tiny phenolic spheres with densities from 10 to 40 pounds per cubic foot are mixed with binder resin to form *syntactic* foam. This foam is used in sandwich construction as in boat hulls and decks and airplane structures. *Urea-formaldehyde* open-cell foams are similar to foamed phenolics.

Cellular Cellulose Acetate

Closed-cell cellulose acetate foam is available in boards and rods. It is used in floats, sandwich construction, and in electronic equipment.

Cellular Polyethylene

Cellular polyethylene, produced by means of a blowing agent, is used mainly for primary insulation on coaxial and other high frequency cables. It has also been used for light-weight spacers, small boat bumpers, and bottle caps.

Other Foams

Epoxy foams are used in applications where the properties of high strength-weight ratio, chemical resistance, and heat resistance up to 250° F are desirable. Several types of foam from *silicone* resins and rubbers are available.

Table 15.2: Fiber Trademark Names, Products, and Manufacturers

Trademark Name*	Product	Manufacturer
Acrilan	acrylic	Chemstrand Corporation
Arnel	triacetate	Celanese Corporation of America
Avisco PE	polyethylene	American Viscose Corporation
Boltathene	polyethylene	Bolta Products Division
Caprolan	nylon	Allied Chemical and Dye Corporation
Cordura	high tension viscose	E. I. du Pont de Nemours & Company
Creslan	acrylic	American Cyanamid Company
Dacron	polyester	E. I. du Pont de Nemours & Company
Darlan	modified saran	B. F. Goodrich Chemical Company
Dynel	modacrylic	Union Carbide Chemicals Company
Fortisan	high tension acetate	Celanese Corporation of America
Fortrel	polyester	Fiber Industries, Inc.
Lycra	spandex	E. I. du Pont de Nemours & Company
Marlex	polyolefins	Phillips Petroleum Company
Orlon	acrylic	E. I. du Pont de Nemours & Company
Prolene	polypropylene	Industrial Rayon Corporation
Velon	saran	Firestone Plastics Company
Verel	modacrylic	Eastman Chemical Products Company
Vyrene	spandex	U.S. Rubber Company
Vycron	polyester	Beaunit Mills
Zefran	acrylic	Dow Chemical Company

* Some fibers are also sold under descriptive or generic names prefixed with the general brand name of the company. Foreign trade names are not listed.

CHAPTER 16

Miscellaneous Plastics

HEAT-RESISTANT PLASTICS

"Heat resistance" is a relative term. The *engineering plastics* show higher resistance than most other thermoplastics. In the space industry high temperature polymers are those which will withstand a minimum of 570° F. Glass-filled phenolics approach this minimum while some silicones exceed 900° F. Several new materials under development show much promise.

Pyrrones are prepared by reacting almost any cyclic dianhydride with an orthotetraamine or derivative. A typical example is made from pyromellitic dianhydride and tetraaminobenzene. The pyrrones may be used at 930° F.

Other heat-resistant polymers under development include polyphenylenetriazoles, polyguinoxalines, polyphenylenetriazoles, polydithizoles, polyamidines, and hydroquinone polyesters.

FURANE RESINS

Furfural (furfuraldehyde) is used in phenol-furfural resins and as an intermediate in nylon 6,6. The term *furane resins* usually refers to products from furfuryl alcohol which is produced by hydrogenation of the aldehyde. The addition of a small amount of acid to the alcohol causes it to polymerize. The completely cured resin has excellent resistance to chemicals and to temperatures as high as 500° F. It is used as a binder for foundry core sands, in laminates, and in chemically resistant coating. With fiber-glass and asbestos reinforcements it is used in industrial products such as tanks, pipe, stacks, ducts, and conduits. Furfuryl alcohol-formaldehyde resins are used as adhesives and impregnants. Furfural-ketone resins are used as coatings and adhesives. Furfurin, produced by reacting the aldehyde with ammonia, can be converted to a hard black solid by the addition of a mineral acid.

COLD MOLDED PLASTICS

These products are molded at room temperatures but must be baked at from 175° to 450° F after molding. They are commonly composed of a bituminous binder, such as asphalt; a drying oil, such as linseed; and a filler, such as asbestos. They are used mainly in electrical parts.

PLASTICS FROM AGRICULTURAL MATERIALS

Considerable research has been done on the utilization of such agricultural by-product materials as corncobs, cornstalks, and the various straws. These are composed chiefly of cellulose, lignin, and pentosans. The pentosans are readily converted to furfural. Thus these residues or by-products are possible raw materials for various plastics. Their utilization for this purpose is limited by economic rather than technical considerations.

Work has been done at Iowa State University on the production of plastics from a reaction of corncobs, sulfuric acid, and phenol. It was hoped that the acid would react with the pentosans to form furfural. This would then react with the phenol to form a phenol-furfural plastic, with the cellulose acting as a filler and the lignin as an extender. The chemistry of the process is more complicated than this since several reactions occur more or less simultaneously. The product was similar to a phenolic plastic but was thermoplastic. Addition of hexamethylenetetramine converted it to a thermosetting product. Another product was made from a mixture of these materials with soybean meal. While these products showed promise they have never been commercialized. Some of the materials, such as corncobs, can, when ground to proper size, be used as fillers in phenolics. Their principal disadvantage is high moisture absorption.

ETHYLENE OXIDE POLYMER

Ethylene oxide is polymerized using strontium carbonate as a catalyst. Polymers have been made with molecular weights above 4 million. The polymer is thermoplastic and can be injected, molded, extruded, and calendered. It is water soluble but can be made insoluble by reacting with a polycarboxylic acid such as polyacrylic. In addition to molded products it is made into water soluble films for special packaging applications. It is also used in textile sizing and in pharmaceutical applications.

Trademark name: Polyox

POLYSPIRALACETAL

This polymer (see Figure 16.1) is made by copolymerizing technical pentaerythritol, a mixture of the mono- and di-compounds, with glutaraldehyde to produce a stable linear polymer that can be cross-linked at the hydroxyl groups. It is expected that it will have use as a wire enamel and as an electrical insulating coating.

RECENT DEVELOPMENTS

Parylene

Parylene is a polymerized paraxylylene available only as a coating. It has superior electrical and heat resistant properties. Probable applications will be in specialized electrical uses.

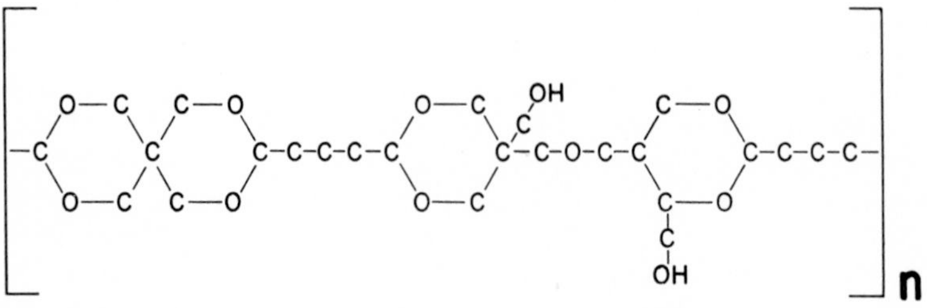

Fig. 16.1—Polyspiroacetal formula.

Noryl

Noryl is based on the PPO technology (see Chapter 13). Uses include water-distribution parts up to 165° F, electrical parts requiring a self-extinguishing material, and appliance and automotive parts where good mechanical properties are needed over a wide temperature range.

FLUOROPLASTICS

Polytetrafluoroethylene, polytrichlorofluoroethylene, and tetrafluoroethylene-hexafluoropropylene copolymer: See under Polyolefins, Chapter 7.

Polyvinylfluoride and polyvinylidene fluoride: See under Vinyl Plastics, Chapter 8.

CHAPTER 17

Some Applications of Plastics

PROPERTIES

The proper application of any material requires a knowledge of the properties of that material. Plastics include such a wide variety of materials that it is difficult to make meaningful generalizations on plastics properties. The problem is complicated by the frequent modification of the basic polymers not only chemically but by the addition of modifying agents such as plasticizers, fillers, extenders, colorants, and reinforcements. The production method, design, and use of the consumer product may also affect the apparent properties of the plastic.

While conventional tests do not form an accurate means of evaluating the performance of plastics under various conditions of use, they do give us comparative guidelines for judgments. Some of the major properties in terms of conventional tests will be considered briefly in a general way and tabulated as ranges of values for the principal plastics materials.

Most of the tests are those made official by the American Society for Testing Materials (ASTM).

Strength of Plastics

Strength may be considered as the resistance of a material to distortion or breakage when subject to mechanical force or

Table 17.1: Some Properties of Plastics*

Plastic	Specific Gravity	Water Absorption	Linear Expansion	Heat Resistance
		per cent in 24 hrs.	*coeff. °C* $\times 10^{-5}$	*(continuous) °F*
Acetals	1.4	0.2–0.9	4.5	185–250
Acrylics	1.2	0.3–0.4	3.0–4.0	160–190
Cellulosics	1.1–1.4	1.0–6.0	8.0–18.0	130–220
Polyesters	1.9	0.1–0.2	1.5–5.0	290–310
Polyethylenes	0.9	0.01	9.0	200–212
Nylons	1.6	0.4–1.5	5.0–8.0	290–310
Phenolics	1.3–1.9	0.1–1.7	1.0–6.0	275–450
Polystyrenes	1.1	0.04–0.5	3.4–21.0	140–200
PVC	1.4	0.5–0.8	5.0–25.0	150–175
Ureas	1.5	0.5–0.7	2.2–3.6	160–180

* Values will vary with specific formulations and additives such as fillers.

stress. This resistance may be measured as tensile, compressive, or impact strength; or in terms of creep, fatigue, and elasticity.

Tensile Strength. Tensile properties are the most important strength characteristics of plastics. *Ultimate tensile* strength is commonly expressed in terms of pounds of pull per square inch of cross-sectional area (ASTM D638). Plastics may vary in tensile strength from 500 psi to as much as 50,000 psi for certain reinforced products. This compares with structural steels with tensile strengths of approximately 40,000 to 100,000 psi. Certain ferrous alloys in fine wire may reach a strength of 250,000 psi.

When a test piece is subjected to tension, elongation occurs. During the first part of the test the elongation, or strain, is reversible and is proportional to the applied load or stress; thus a plot between stress and strain values produces a straight line, as *OA,* in the Figure 17.1. The numerical value of this relationship is referred to as the *elastic modulus* or *tensile modulus.* The point *A* is the *yield stress* above which the relative rate of elongation increases and where permanent distortion occurs. This may take the form of *AB,* characteristic of polyethylene, or of *AC,* characteristic of dry nylon. A third type of failure is by fracture before yielding as *AD,* characteristic of acrylics. The elastic modulus is a more important characteristic from the utilization standpoint than ultimate tensile strength.

Plastics fibers such as nylon, rayon, or polypropylene, when stretched, increase in strength. This results from orientation of the molecular chains so that they lie parallel along the length of the fiber. As the chains are drawn closer together the attractive forces increase, resulting in increased resistance to longitudinal

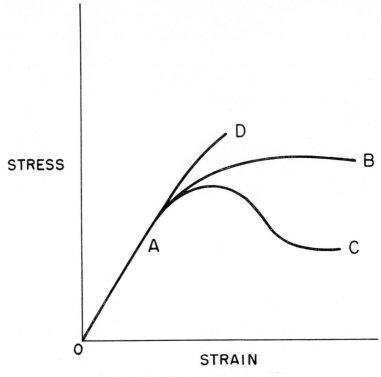

STRESS

STRAIN

Fig. 17.1—Stress-strain curves.

slippage. Films may be stretched in one direction or in two directions to give increased strength. In injection molding of
polypropylene it is possible to orient the fibers so that built-in
hinges can be produced which will stand almost infinite flexing.

Materials subjected to stresses in the elastic part of the curve
below the yield point may gradually elongate beyond the initial
elastic deformation. With the release of the load only the elastic
part of the stretch is reversed, leaving some permanent distortion.
This phenomenon known as *creep* varies with the variety of
plastic and the temperature. Repeated application and removal
or reversal of a stress in the elastic area may result in failure by
fatigue. Fatigue breakage occurs more readily as the yield point
is approached. Tests involve the application of reversible stresses
over a large number of cycles.

Compressive Strength. Compressive strength (ASTM D695)
alone, except in expanded materials (foams), is usually not a

critical characteristic in plastics. When pressure is applied at midpoint of a rectangular or beam-shaped sample supported horizontally on two rounded wedges, the sample is stressed in compression in the top one-third and in tension in the bottom one-third. Loading at failure is known as the flexural strength (ASTM D790). For materials that do not break, the results may be reported as the loading in pounds per square inch necessary to stretch the outer surface 5 per cent. For stiffness *in flexure* see ASTM D747.

Impact Strength. In many applications of plastics the impact strength is important. In the *Izod impact* test (ASTM D256) a notched sample is broken by a swinging pendulum and the energy required measured by the length of the pendulum swing. Data for various plastics are given in Table 17.2.

Thermal Effects

True thermoplasts soften when heated and may even become liquid. True thermosets do not soften with heat. Between the two are materials which soften in varying degrees when heated. Deformation under load may increase as the temperature rises. This deformation is sometimes defined in terms of the temperature at which a 66 or 264 psi fiber stress applied to a sample supported as a beam will deflect 0.01 inch (ASTM D648).

In general the expansion of plastics by heat is greater than that of metals and may vary from 0.5×10^{-5} per °F for thermosets and up to 13×10^{-5} per °F for some thermoplasts. This may be an important factor in the design of parts combining plastics and metals which may be subject to considerable temperature

Table 17.2:   Strength of Plastics*

Plastic	Tensile	Flexural	Impact	Compression
	psi	*psi*	*ft. lb. per in.*	*psi*
Acetals	8,800–10,000	13,000–14,000	1.2–1.8	16,000–18,000
Acrylics	3,000–11,000	8,000–19,000	0.4–10.0	4,000–19,000
Polycarbonates	9,500–18,000	13,500–36,000	5.0–14.0	12,500–18,000
Cellulosics	2,000–11,000	2,000–16,000	0.4–10.0	2,200– 3,500
Polyesters	3,000–55,000	6,000–80,000	0.2–24.0	15,000–50,000
Polyethylenes	1,000– 4,400	1,000– 7,000	0–23.0	3,200
Polystyrenes	2,000– 8,500	5,000–14,000	0.3–8.0	4,000–16,000
PVC	1,000– 9,000	10,000–16,000	0.5–10.0	900–13,000
Phenolics	5,000– 9,000	8,000–12,000	0.3–0.9	20,000–35,000

* Typical figures from manufacturers' literature.
Note: Different fillers produce variations in strength.

changes. The *specific heat,* or heat capacity, for most plastic resins is between 0.2 and 0.7 BTU per lb. °F.

Electrical Characteristics

In general plastics may be considered as nonconductors of electricity. The following electrical properties are commonly reported. *Volume resistivity* (ASTM D257) is the ohmic resistance of the bulk material. *Surface resistivity* is the resistance between electrodes on the plastic surface. *Dielectric strength* (ASTM D149) is the voltage per unit of thickness required to break down the material. *Dielectric constant* (ASTM D150) is the ratio of capacities of a condenser using the plastic and one using air as the dielectric. (See Table 17.3.)

Resistance to Chemicals

The resistance to moisture and chemicals varies greatly among various plastics. Moisture absorption is usually expressed as the per cent water absorption on immersion for 24 hours of a piece ⅛-inch thick (ASTM D570). While chemical resistance in some cases might be expressed as loss or gain in weight, general qualitative terms are usually used. For example, the "effect of weak acids" may be expressed as "resistant" or "slight."

Table 17.3: Electrical Properties*

Plastic	Dielectric Constant	Dielectric Strength	Volume Resistivity	Power Factor (Dissipation)
	60 cyc. D150†	*⅛″ D149*†	*23° C R257*†	*60 cyc. D150*†
Acetals	3.7	500	6×10^{14}	0.003
Acrylics	3.0–3.9	450–500	$>10^{14}$	0.04–0.06
Cellulose Acetates	3.5–4.5	250–600	$10^{10}–10^{15}$	0.01–0.06
Cellulose Nitrates	7.0–7.5	300–600	$10–15(10^{10})$	0.09–0.12
Chlorinated Polyethers	3.44	400	3.1	0.011
Nylons	3.4–4.0	Excellent	$10^{14}–10^{15}$	0.014–0.19
Phenolics	5.0–5.5	180–450	$10^{11}–10^{15}$	0.03–50.0
Polycarbonates	3.16	400	$2.1–10^{16}$	0.0009
Polyesters	3.0–7.5	350–500	$10^{12}–10^{15}$	0.01–0.04
Polyethylenes	2.3–2.6	500–700	6×10^{14}	0.0002–0.005
Polypropylenes	2.2–3.4	450–500	$10^{15}–10^{17}$	0.0001–0.007
Polystyrenes	2.5–2.75	400–600	10^{14}	0.009–0.012
PVC	3.3–11.7	340–400	$10^{11}–10^{16}$	0.051–0.103
Silicones	5.0	160	0–1000	0.015
Ureas	7.6–7.8	300–390	$10^{12}–10^{13}$	0.35–0.43

* From manufacturers' literature.
† ASTM test number.

Table 17.4: Chemical Resistance of Plastics*

Plastic	Acids		Alcohols	Alkalies	Ethers and Ketones	Hydrocarbons			Fats and Oils	Inorg. Salts
	HCl and H_2SO_4	HNO_3†				Aliphatic	Aromatic	Chlorinated		
Acetals	U	U	E	U–E	E	E	E	E	E	E
Acrylics	E	U	U	F	U	U–E	U	U	E	G
Polycarbonates	G–E	G–E	E	U–E	U	E	U–E	U	E	E
Cellulosics	U	U	F	U	U	E	U	U	E	E
Polyesters	G	G	G	F	U	E	E	G	E	E
Polyethylenes	G–E	U–E	E	E	U	U	U	U	E	E
Chlorinated Polyethers	A	U–E	E	E	G–E	E	G	G	E	E
Nylons	U	U	E	E	G	E	E	G	E	⋮
Phenolics	G	U	U	U–G	F	E	G	G	G	E
Polystyrenes	E	U	G	E	U	F	U	U	E	E
PVC	E	E	E	E	U	E	U	U	E	E

* From manufacturers' statements offered as general guides which should be checked for specific applications. E = excellent; G = good; F= fair; U = unsuitable. (For "room temperatures.")
† In general for other oxidizing acids and for hypochlorites.

INDUSTRIAL EQUIPMENT

Plastics are frequently used as materials for the construction of industrial equipment because of their resistance to action of chemicals which might corrode other materials of construction such as metals. This resistance to chemicals has been noted in a general way in the description of the individual plastics. It may vary with degree of polymerization; amount and kind of added plasticizers, fillers, or special additives; and temperature. Strains incident to molding or fabrication may make the material more sensitive to certain chemical solutions, resulting in phenomena such as stress cracking of polyethylene. Before final adoption of a plastics material for use it is well to consult the manufacturer for any available specific data relating to the proposed use. Frequently laboratory tests under simulated use conditions are desirable.

Mechanical strength needs careful appraisal before deciding upon an application. In general the factors affecting chemical resistance may also affect mechanical strength. Methods of molding or fabrication may be important. It is common knowledge that plastics, particularly thermoplastics, will show creep under continuous loads. The tensile strength and related values under load may decrease to a permanent value. This makes it desirable to know the relation between short-term strength values and final values resulting under continuous loading. Notching a plastic piece decreases its impact strength, hence it is common to report impact strength values secured on notched samples.

Plastics are light in weight compared with other structural materials, varying from two to 130 pounds per cubic foot for a series from cellular products to highly filled material. Thus the plastic product is frequently lighter and more easily supported than a similar product from another material, especially steel. The strength-weight ratio is usually favorable. Thermal conductivity, in general, is low. This may result in the need for less insulation. The thermal expansion of plastics is higher than that of other structural material. This must be considered when plastics are combined with metals. In general, plastics are not suitable for use at as high temperatures as such inorganic materials as steel, glass, or concrete. However, as will be pointed out later, some plastics retain their usefulness at remarkably high temperatures. It must be remembered that plastics include a wide variety of materials with a wide range of properties. These generalizations merely point out the need for careful design. Details of design are beyond the scope of this book.

Cost of equipment depends both upon the cost of the raw

plastics material and the cost of fabricating. The latter will vary with the detailed method if more than one is available. If only one piece is to be made, welding from sheets or other standard shapes or molding by hand lay-up may be the cheapest fabricating method. If quantity justifies expensive molds or matched dies the use of these may be more economical. In considering the cost of the equipment, service life must be considered. Frequently the longer life of plastic equipment over that made from some cheaper material such as steel is enough greater to justify a much higher first cost. Cost comparisons need to be made among plastics as well as with other materials.

The applications of plastics are continually being expanded and new products and modifications of old ones are being developed. Some of the most interesting applications have only very limited use. An attempt will be made in this chapter to cover some of the more common and important applications. The reader will find other applications described in technical and trade journals and in manufacturers' descriptive material.

Pipes, Fittings, and Valves

Piping and tubing are made from a variety of plastics and thus are available with a considerable range of physical properties. In general, plastic pipe is used because it is superior to common metals in corrosion resistance. The corrosion resistance of the pipe is, of course, basically that of the plastic from which it is made. Plastic pipe has to be used with due regard to temperature, pressure, and mechanical strength limitations.

Polyvinyl Chloride Pipe and Fittings. Polyvinyl chloride pipe is produced by extrusion in two types: normal-impact and high-impact. The normal-impact pipe has better chemical resistance and can be used at higher pressures than the high-impact pipe. The high-impact pipe is made from a copolymer and has greater resistance to shock loads and rough handling than the normal-impact product. Pipes with two wall thicknesses, "Schedule 40" and "Schedule 80," are the common weights. "Schedule 120," an extra heavy pipe, is also standard. As with metal pipe the thickness of the wall varies with the pipe diameter. For 1-inch nominal pipe the wall thicknesses for Schedules 40, 80, and 120 are 0.133, 0.179, and 0.200, respectively.

Injection-molded fittings of both normal- and high-impact polyvinyl chloride are available. These include such standard fittings as elbows, tees, caps, couplings, unions, reducing bushings, plugs, and flanges. These are very similar to standard iron pipe fittings and are made in two types: threaded and socket. The socket fittings slip over the end of the pipe and may be solvent cemented to produce a strong joint. Threaded fittings are not recommended for Schedule 40 pipe because of the notch sensitivity of polyvinyl chloride which results in a weak joint. Thermosetting resin types of adhesives cured by a catalyst can also be used with the socket fittings. The pipe can be welded using hot air or inert gas with suitable filler rods.

Recommended maximum operating pressures at 75° F for Schedule 40 normal-impact polyvinyl chloride pipe, with solvent cemented fittings, varies from 125 psi for 6-inch pipe to 410 psi for ½-inch pipe. These pressures drop a little under one-half at 150° F. The recommended spacing of supports varies from 5.0 to 8.5 feet at room temperature to 2.0 to 4.0 feet at 150° F, maximum recommended temperature.

Polyvinyl chloride pipe is used for water lines, sour crude oil lines, salt water disposal, and in chemical plant piping for handling a variety of chemicals including acids, alkalies, salt solutions, and alcohols. Electrical conduit and fittings made from both polyethylene and polyvinyl chloride (see Figure 17.2) have the advantage of good electrical insulation and corrosion resistance.

Hard Rubber Pipe. Hard rubber pipe is produced in sizes from ¼ inch to 4 inches. Fittings corresponding to standard threaded and flanged iron pipe fittings are available. Hard rubber pipe in standard weight is not recommended for use with pressures over 50 psi at room temperature nor at temperatures above 120° F. It is used for handling acids and chemical solutions.

Saran Pipe. Extruded saran pipe is available in a wall thickness corresponding to Schedule 80 in sizes from ½ inch to 6 inches. Threaded molded fittings corresponding to iron pipe fittings are available up to the 2-inch size. Flanges and reducing flanges are made up to the 6-inch size. Fittings from 2½- to 6-inch sizes are fabricated by welding suitable shapes cut from pipe.

Saran pipe has excellent resistance to most chemicals at room temperature. Exceptions include ammonium hydroxide, liquid halogens, and some aromatic ketones. Resistance to chemicals decreases at higher temperatures, particularly above 125° F. Working pressures vary from 60 psi for 6-inch to 260 psi for ½-

Fig. 17.2—Rigid polyvinyl chloride conduit. (Courtesy B. F. Goodrich Chemical Company)

inch pipe. Top temperature for use under zero stress conditions is 170° F.

ABS Pipe. ABS (acrylonitrile-butadiene-styrene) pipe is made by extrusion in Schedules 40 and 80 in ½-inch to 2-inch sizes and in a thinner-walled product up to 6 inches. Injection-molded screwed fittings are available. The 1-inch Schedule 40 pipe is given a working pressure of 260 psi at 75° F. Maximum temperature for satisfactory use is 180° F. Applications are similar to those for polyvinyl chloride pipe.

Polyethylene Pipe. This pipe is extruded in sizes up to about 8 inches (see Figure 17.3). The Schedule 40 weight is common but a heavier pipe corresponding to Schedule 80 is also made. For sizes one foot or more in diameter it can be produced by centrifugal casting using a revolving metal pipe as a mold. Pipe from low-density polyethylene up to 2 inches in diameter is sold in

Fig. 17.3—Polyethylene pipe. (Courtesy Hercules Powder Company)

coils commonly 200 feet long. High-density pipe under one inch is available in coils, one inch and above in 20-foot lengths. Centrifugally cast pipe is in 10-foot lengths. Insert fittings from high impact polystyrene, butyrate, or brass held in place with stainless steel clamps are commonly used. Polyethylene pipe for outdoor applications is made resistant to ultraviolet light by the addition of well-dispersed carbon black and an antioxidant. Polyethylene pipe is approved for handling drinking water and food products. Polypropylene pipe is similar to polyethylene pipe but can be used up to about 150° F compared to 120° F for the polyethylene.

Among the advantages of polyethylene and polypropylene pipes are excellent resistance to a wide variety of solvents and solutions of alkalies, acids, and salt; freedom from taste, odor, and toxicity; good flexibility and toughness; low resistance to flow; light weight; and low cost. Horizontal runs of the pipe must be supported at close enough intervals to prevent sagging. Because of its flexibility and smooth surface it is frequently possible to pull polyethylene pipe through walls or old metal pipes.

Cellulose Acetate-Butyrate Pipe. This pipe is made by extrusion in sizes up to 6 inches and in at least two wall thicknesses. Threaded and slip joint injection fittings are available. Large-sized fittings can be fabricated by cutting and cementing techniques. The pipe is readily cemented. Cellulose acetate-butyrate pipe is used for handling salt water, crude oil, natural gas, wine, beer, fruit juices, and various chemicals.

Glass Fiber-Reinforced Pipe. Pipe is manufactured from polyesters, epoxies, and furane resins reinforced with glass fiber. The polyester pipe has been extruded. Centrifugal casting inside tubing has also been employed. Several methods of wrapping glass fiber around a mandrel and saturating it with the resin are used for polyester, epoxy, and furane resin pipe. One method is to extrude the resin onto a mandrel, followed by wrapping with resin-saturated glass fiber. The glass fiber may be in the form of cloth, tape, or roving. Automatic equipment for continuously making pipe is in use. Pipe as large as six feet in diameter has been produced. The glass-reinforced pipes in general are resistant to many chemicals and can be used at higher temperatures and pressures than most of the other plastic pipe. It is claimed that furane pipe can be used up to 300° F and up to 150 psi. Epoxy pipes, especially those produced by spiral winding of glass filaments or tape, also show excellent temperature and pressure resistance. Working pressures as high as 1000 psi are claimed.

Glass-reinforced pipe can be damaged by careless handling, causing a separation of the inner resin lining, exposing the glass fibers. Hammering or surging in a pipe in use may also loosen the resin. When handling solids, the pipe may be subject to abrasive wear. Leaks are readily repaired by patching with glass cloth and resin without removing the pipe.

Glass fiber-reinforced polytetrafluoroethylene pipe has a useful temperature range of —100° F to +500° F and is inert to practically all chemicals. Resistance to thermal shock, physical shock, strain, and vibrations is excellent.

In addition to high-pressure and high-temperature applications, reinforced plastic pipe shows promise of use in areas such as acid-resistant sewers. A one-mile 3¼-foot diameter reinforced plastic pipe is in use in Sweden carrying acid waste water.

Phenolic Plastic Pipe. Pipe from phenol-formaldehyde resin reinforced with asbestos fiber is made in sizes from ½ inch. Flanged fittings are used. The pipe has excellent resistance to hot hydrochloric acid as well as many other chemicals. It may be used up to 300° F and at pressures from 30 to 65 psi.

Plastic-lined Steel Pipe. Saran-lined pipe and flanged fittings are available in sizes of 1 inch through 8 inches. This has the advantage of the chemical resistance of the saran with the supporting strength and stiffness of the steel. Operating temperatures range from —20° F to 200° F. Maximum pressure rating at 200° F is 300 psi.

Pipe with both soft and hard rubber linings in sizes of 2 inches through 12 inches is available. Both types can be used up to 185° F and to working pressures of either 125 psi or 250 psi. Flanged fittings are available.

Polytetrafluoroethylene-lined steel pipe is also available. It can be used up to 500° F. Pipe lined with chlorinated polyether has recently become available. It has excellent chemical resistance.

Valves. Valves are made from various plastics. Globe valves are available in polyvinyl chloride and polyethylene in sizes of ½ inch through 2 inches. Similar valves known as Y-valves are available in phenolic compositions in sizes of ½ inch to 10 inches. Check valves and ball valves are made in polyvinyl chloride, ABS, polypropylene, hard rubber, and chlorinated polyether in ½-inch to 4-inch sizes. Phenolic check valves are available up to 10 inches. Plastic-lined metal valves are also available. Hard rubber gate valves are available up to 20 inches. Sampling valves molded from phenolic resin are available for use on filter presses (see Figure 17.4). Acetals and polycarbonate are also coming into use as valve material.

OTHER PROCESSING EQUIPMENT

Pumps and Fans

Centrifugal pumps and fans are made of phenolic resin (see Figure 17.5), polycarbonate, polyethylene, acetals, polyvinyl chloride, chlorinated polyether, hard rubber, and glass-reinforced polyesters and epoxies. The housings from polyethylene, polyvinyl chloride, and chlorinated polyether may be fabricated by machining and joining together of parts from sheet material by welding or mechanical fastenings such as bolts. Hard rubber parts can also be bolted together. Bolts machined from hard rubber and polyvinyl chloride are available. Housings from glass-reinforced plastics, phenolics, hard rubber, and chlorinated poly-

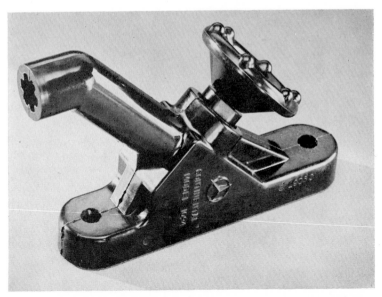

Fig. 17.4—Phenolic filter press sampling valve. (Courtesy Durez Plastics, Hooker Chemical Corporation)

Fig. 17.5—Phenolic pump impeller. (Courtesy Durez Plastics Division, Hooker Chemical Corporation)

ether, particularly in the smaller sizes, may be molded. Impellers and other parts may be of molded or of welded construction. Stainless steel or other corrosion-resistant metal is commonly used for shafts. These may be further protected by a plastic coating. Pumps and fans are also constructed with steel housings and impellers coated with plastics. Small gear pumps may be constructed of hard rubber or other plastic materials.

Tanks

Plastic tanks are made in a wide range of sizes and shapes and of several materials. Furfuryl alcohol resins reinforced by fiber glass are fabricated over simple forms without pressure to make round tanks or "jars" in 1,200-gallon and smaller sizes. Rectangular and other shapes can also be made. Phenolic resins with asbestos or graphite can be formed into cylindrical tanks up to 10 feet in diameter and 12 feet deep. The larger sizes are reinforced on the outside by vertical wood staves and metal hoops. Flat-, dished-, and cone-bottomed tanks are standard. Rectangular tanks are also made.

Polyvinyl chloride and polyethylene tanks are made in a wide variety of sizes and shapes from sheets by welding. Polyethylene drums up to 50-gallon capacity are made by blow-molding.

Glass-reinforced polyester and epoxy resin tanks are produced in cylindrical forms by low pressure lay-up methods and by filament winding. For example, tanks made by the latter method 12 feet in diameter and 20 feet high are used for storage of sour crude. Tanks on tank trucks handle a variety of chemicals as well as food products such as milk and fruit juices. For example, high-test sodium hypochlorite solution is being handled in 4,300-gallon glass-reinforced polyester tank trailers. It is estimated that around 2,000 trucks with the main tanks or bodies of reinforced plastics are in use. As many as 3,000 more probably have cabs or other large sections made of plastic. Glass-reinforced plastic tanks both rectangular and cylindrical are used in factories.

Steel tanks may be lined by plastic sheets, including natural and synthetic rubber, chlorinated polyether, saran, and chlorosulfonated polyethylene. Other plastics may be applied by spraying in molten or solution form.

Miscellaneous Chemical Processing Equipment

Plastics are becoming increasingly important in this area not only for tanks but many other pieces of equipment. A few will be mentioned. Various types of towers, such as packed, tube, bubble cap, sieve plate, and cascade, are made from phenolic resins with asbestos and graphite fillers and from glass-reinforced polyester resins. Typical of these are packed towers of polyester resins up to 42 inches in diameter and 18 feet high, handling such chemicals as chlorine; hydrofluoric and nitric acid mixture; hydrogen fluoride, silicon fluoride, and sulfur dioxide combination; and hydrochloric acid.

Polyvinyl chloride and polyethylene can be fabricated by welding into a wide variety of chemical processing equipment. Typical polyethylene applications include conveyor chutes for coal and corrosive chemicals, dipping baskets for metal pickling, safety jugs, bottle carriers, and fume scrubbers for hydrochloric acid. Similar equipment is made from glass-reinforced polyester and furane resins. Other equipment is made from plastic-coated steel, including such sheet coatings as rubber, saran, polyvinyl chloride, and chlorinated polyether.

Hoods, ventilating ducts, and vent stacks for handling acid fumes and other corrosive materials are made from the asbestos and graphite-filled phenolics, from welded polyvinyl chloride and polyethylene sheets, and from glass-reinforced polyesters and epoxies (see Figure 17.6). A stack 150 feet high and 72 inches in diameter made of glass-reinforced polyester is in use handling

Fig. 17.6—Glass-reinforced polyester ventilating ducts. (Courtesy Hooker Chemical Corporation)

acid fumes from a chemical plant. This stack was constructed in 24-foot sections. Glass-reinforced furfuryl alcohol resins are used in towers available in standard sizes from 6 inches to 36 inches in diameter and in 10-foot sections.

Two interesting plastics applications in filtration may be cited. One is a 100-gallon Buechner funnel made of high-density polyethylene. The other is the production from asbestos-filled phenolic molding compound of filter plates for plate-and-frame filter presses. These have been made up to 24 square feet in area and weighing 285 pounds by compression molding in a 200-ton hydraulic press.

Another interesting application is the use of polyvinyl chloride pipe for the rollers and shafting in a roller conveyor system which is one-half mile in total length and handles the charging and refilling of batteries with acid. Rollers are of 1.5-inch polyvinyl chloride pipe on axles of ¼-inch Schedule 80 pipe with bushings at the roller ends. Acetal plates are also in use as attachments to roller chains for conveying.

A suggested application is the use of thin plastic film such as polyester for heat exchangers. Since the film will not stand much pressure, horizontal tube-type exchangers operating at low velocities with the same fluid levels on both sides of the tubes might work best.

SHIPPING CONTAINERS

Bottles and Drums

Bottles in a wide range of sizes and shapes are blow-molded from polyethylene. Various large bottles or carboys up to 15 gallons in capacity are used with protective overpacks of corrugated paper board, wood, or steel. A composite package made up of a cube-shaped polyethylene insert in a protective wire-bound box is made in four sizes up to 15 gallons. Drums and drumliners for steel drums up to 55-gallon capacity are blow-molded.

"Dracones"

Dracones are "flexible barges" used for transporting heating oil and similar products. A typical one with a capacity of 10,000 imperial gallons of heating oil is a tube 4.75 feet in diameter and 100 feet long with a tapered end. It is constructed of nylon, coated externally with polychloroprene (Neoprene) rubber and lined with nitrile rubber. The loaded dracone floats in water and is towed by a tug. When empty, the tube is rolled up and carried on the deck of the tug.

"Baggage Expediter"

A new development designed to speed up the loading and unloading of baggage from jet airliners has been put into use (see Figure 17.7). This "baggage expediter" is molded from glass-reinforced polyester resin. Six units, each holding up to 35 bags, are carried in the forward compartment of a jet airliner. This method of handling will doubtless be extended to other objects in other transportation fields.

Miscellaneous

Polyethylene and other plastic bags have been used for a wide variety of packaging. They are also used as liners in paper or cloth bags. Large plastic bags have been used for transport-

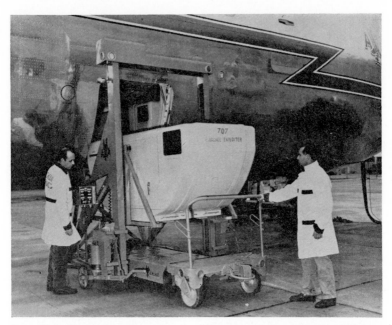

Fig. 17.7—"Baggage expediter." (Courtesy Durez Plastics Division, Hooker Chemical Corporation)

ing granular and powdered products being loaded full onto railway cars. Disposable polyethylene liners are used in these large collapsible plastic containers to haul liquids such as milk, fruit juices, vegetable oils, and other products which must be kept free of contamination. The containers, which come in 22-, 30-, and 34-foot lengths with capacities up to 4,600 gallons, are transported in railway cars or trucks. When the containers are to be used for dry products the polyethylene liner is removed and discarded. Reinforced epoxy plastic containers constructed over a heavy cardboard form and reinforced by a steel frame have been built for shipment of dry resins. One of these is 24 feet long by 8 feet by 8 feet. Large containers for dry ice shipment are made of reinforced polyester inner and outer skins with polyester foam between.

HIGH TEMPERATURE APPLICATIONS

Most plastics are basically high organic polymers and are not suited for high temperature applications. As has been pointed out (Chapter 1) substitution in the polymers with chlorine and fluorine may increase their resistance to heat. The complete substitution of the hydrogens in polyethylene increases the continuous heat temperature from 220° F to 500° F. The high silicon content of the silicone resins (Chapter 12) makes it possible for them to withstand as much as 600° F. However, most commercial plastics are not suitable for continuous use above 500° F. The principal interest in high temperature plastics is for use in missiles, particularly in the nose cones.

Short-Time High Temperature Applications

During reentry into the earth's atmosphere the nose cone of a rocket heats up to between 20,000° and 30,000° F for about 20 seconds. It has been found that certain phenolics and phenyl silanes could be used in this application because of the phenomenon of *ablation*. Ablation in rocketry refers to the disintegration or "erosion" of the hot outer side of the cone while enough of the inner part remains undamaged to provide necessary structural strength. Usually glass, asbestos, or quartz fibers are used to reinforce the organic binder. As the temperature rises, some additional cross-linking of the polymer probably occurs first. Then the outer layer is broken down by the heat and flakes off. The exposed fibers melt to form a highly reflective surface which resists further erosion and reflects the heat away from the surface. The low conductivity of the resin protects the instruments inside the cone.

Long-Time High Temperature Plastics

Long-time exposure in the *blast tube,* which carries thrust gases from the combustion chamber to the exit cone, may only be several minutes, but temperatures may be around 5,000° to 8,000° F. Metals, because of their high conductivity, melt before they reach these temperatures. Here again phenolic resins are used.

For longer times, ablation will not solve the problem. For long-time service the best commercial plastics are in the 500° to 550° F range. In intermittent service these can be boosted to 600° to 800° F. These are mainly tetrafluoroethylene, triallyl cyanurate cross-linked polyester, silicone resin, phenyl silane, and new epoxies.

Research studies have shown some interesting possibilities.

Perfluoroalkyl amidine polymers stable up to 750° F have been synthesized. Some "inorganic polymers" show promise. It is reported that a dicyclohexyphosphinoborine trimer has been made which is stable at 900° F. Polyaluminosiloxanes stable also at 900° F have been reported from Russia. Polymers with similar stability are reported using alternating aluminum and oxygen atoms on the polymer backbone. A variety of other products, including tin- and silicone-containing polymers, are under study.

BUILDING APPLICATIONS

Conventional Construction

Plastic products are used in ordinary building construction in a variety of functional and decorative applications. Cellular products, such as polystyrene, are used for heat insulation. Translucent sheets such as ureas, acrylics, polystyrenes, and polyesters are used as ceilings under lights or in panel construction. Corrugated glass-reinforced polyester sheets have been used as curtain walls in industrial buildings. They admit considerable light without undesirable glare and are resistant to many chemical fumes. Acrylic, polycarbonate, and reinforced polyester panels have replaced glass in windows since they are less readily damaged by storms and vandalism.

Reinforced clear PVC has been used in panels in home construction. The opaque, rigid material has also been used for wall panels, window sash, siding, doors, and eaves and downspouting. Sandwich panels using cellular insulating plastic with vinyl or glass-reinforced polyester surfaces are used for wall construction. Flooring materials include vinyl and rubber tile, epoxy terrazzo, and monolithic-poured polyurethane.

Entire bathroom units have been molded from reinforced polyester and built up from vinyl sheets. Polyester bathtubs and lavatories thermoformed from plastic sheets are available. Water pipes of polyethylene, polypropylene, ABS, and PVC are used.

Dome Type Roofs

Acrylic sheets and panes are used in canopy and dome roof applications (see Figure 17.8). One example is the use of curved 5- by 12-foot sheets as a transparent cover over a patio dining area of a large hotel. Another is the use of 350,000 pounds of

Fig. 17.8—Swimming pool enclosure glazed with acrylic sheets. (Courtesy Rohm and Haas Company)

cast acrylic panes in the dome of the Houston, Texas, stadium popularly called the "Astrodome."

Radomes

Radomes are large spheres used to house radar antennas. Plastic radomes are of interest because they indicate possible construction techniques for industrial buildings of the future. Two general types have been built: rigid structures and inflatable structures.

A rigid radome 140 feet in diameter has been constructed of panels with kraft paper honeycomb core coated with glass-reinforced fire-retardant polyester resin. Typical panels were hexagonal, weighed 125 pounds, and had 31 square feet of area. The panels were bolted together to form a rigid structure.

Inflatable radomes and other structures have been made from plastic sheet material. These are held up by maintaining a slightly higher air pressure inside than out. One of these radomes used several years in Alaska was about 60 feet in diameter and made of polyester fabric with a synthetic rubber ("Hypalon") coating.

Future Uses

The applications cited point out only part of the plastic products that enter into building construction. Uses are increasing but have been slowed down somewhat by building code restrictions. Improvements in quality, especially in the area of better fire resistance, will eventually change the codes. Some opposition is from the building trade unions who fear less work available than with older, more conventional materials.

Appendix

Trademark Name	Product	Manufacturer
Acrylite	acrylic	American Cyanamid Company
Adiprene	urethane rubber	E. I. du Pont de Nemours & Company
Agilene	polyethylene	American Agile Corporation
Agilide	PVC	American Agile Corporation
Alathon	polyethylene	E. I. du Pont de Nemours & Company
Alkathene	polyethylene	Imperial Chemical Industries, Ltd.
Ameripol	synthetic rubber*	Goodrich-Gulf Chemicals
Ampacet	polyethylene	American Molding Powder & Chemical Corporation
Arodure	urea resin	Archer Daniels Midland Company
Arothane	urethane foam	Archer Daniels Midland Company
Bakelite†	various plastics	Union Carbide Corporation
Beetle	urea-formaldehyde	American Cyanamid Company
Bexoid	cellulose acetate	B. X. Plastics, Ltd.
Blacar	polyvinyl chloride	Cary Chemicals, Inc.
Butacite	polyvinyl acetal	E. I. du Pont de Nemours & Company
Butvar	polyvinyl butyral	Shawinigan Resins Corporation
Celcon	acetal	Celanese Corporation of America
Chemigum	synthetic rubber*	Goodyear Tire & Rubber Company
Cumar	coumarone-indene resin	Allied Chemical Corporation

199

Trademark Name	Product	Manufacturer
Cycolac	ABS	Marbon Chemical Division, Borg-Warner Corporation
Cycolon	ABS	Marbon Chemical Division, Borg-Warner Corporation
Cymel	melamine-formaldehyde	American Cyanamid Company
Dapon	diallyl phthalate	FMC Corporation
Delrin	acetal	E. I. du Pont de Nemours & Company
Durethene	polyethylene film	Sinclair-Koppers Company, Inc.
Durez†	various plastics	Durez Division, Hooker Chemical Corporation
Dyalon	urethane elastomer	Thombert, Inc.
Dylan	polyethylene	Sinclair-Koppers Company, Inc.
Dylene	polystyrene	Sinclair-Koppers Company, Inc.
Dylite	expandable polystyrene	Sinclair-Koppers Company, Inc.
El Rex†	various plastics	Rexall Chemical Company
Elvanol	polyvinyl alcohol	E. I. du Pont de Nemours & Company
Elvax	vinyl resin	E. I. du Pont de Nemours & Company
Epi-Rez	epoxy resin	Davoe & Raynolds Company
Epolite	epoxy resin	Rezolin, Inc.
Epon	epoxy resin	Shell Chemical Company
Escon	polypropylene	Enjay Chemical Company
Estane	polyurethane	B. F. Goodrich Chemical Company
Ethafoam	expanded polyethylene	Dow Chemical Company
Ethocel	ethyl cellulose	Dow Chemical Company
Exon	polyvinyl chloride	Firestone Plastics Company
Formvar	polyvinyl formal	Shawinigan Resins Corporation
Forticel	cellulose propionate	Celanese Corporation of America
Fortiflex	polyethylene	Celanese Corporation of America
Fostalite	polystyrene	Foster Grant Company, Inc.
Fostarene	polystyrene	Foster Grant Company, Inc.
Gelvatol	polyvinyl alcohol	Shawinigan Resins Corporation
Genpol	polyester	General Tire & Rubber Company
Genthane	polyurethane rubber	General Tire & Rubber Company
Geon	PVC	B. F. Goodrich Chemical Company
Glaskyd	glass-reinforced alkyd	American Cyanamid Company
Halon	fluorohalocarbon	Allied Chemical Corporation
Hetrofoam	fire-retardant polyurethane	Durez Division, Hooker Chemical Corporation
Hetron	fire-retardant polyesters	Durez Division, Hooker Chemical Corporation
Hi-fax	polyethylene	Hercules Powder Company
Hypalon	chlorosulfonated polyethylene	E. I. du Pont de Nemours & Company
Insular	polyvinyl chloride	Rubber Corporation of America
Irrathene	irradiated polyethylene	General Electric Company
Kel-F	poly CTFE	Minnesota Mining & Manufacturing Company
Kralastic	ABS	U.S. Rubber Company

Trademark Name	Product	Manufacturer
Kynar	polyvinylidene fluoride	Pennsalt Chemicals Corporation
Laminac	polyester	American Cyanamid Company
Lexan	polycarbonate	General Electric Company
Linde	silicones	Union Carbide Corporation
Lucite	acrylic	E. I. du Pont de Nemours & Company
Lustran	ABS	Monsanto Company
Lustrex	polystyrene	Monsanto Company
Maraset	epoxy resin	Marblette Corporation
Marlex	polyolefins	Phillips Petroleum Company
Marvinol	PVC	U.S. Rubber Company
Merlon	polycarbonate	Mobay Chemical Company
Moplefan	polypropylene	Montecatini, Società Generale per l'Industria Mineraria e Chemiche
Mouldrite	phenolic and urea	Imperial Chemical Industries, Ltd.
Mylar	polyester film	E. I. du Pont de Nemours & Company
Nordel	ethylene-propylene rubber	E. I. du Pont de Nemours & Company
Oleform	polypropylene	Avisun Corporation
Olemer	propylene copolymer	Avisun Corporation
Opalon	PVC	Monsanto Company
Pelaspan	expandable polystyrene	Dow Chemical Company
Penton	chlorinated polyether	Hercules, Inc.
Petrothene	polyethylene	U.S. Industrial Chemicals Company
Plaskon†	various plastics	Allied Chemical Corporation
Plastacele	cellulose acetate	E. I. du Pont de Nemours & Company
Plexiglas	acrylic	Rohm and Haas Company
Pliovac	PVC	Goodyear Tire & Rubber Company
Plyophen	phenolic	Reichhold Chemicals, Inc.
Poly-Eth	polyethylene	Spencer Chemical Company
Polylite	polyester	Reichhold Chemicals, Inc.
Pro-fax	polypropylene	Hercules, Inc.
Resimene	melamine-formaldehyde	Monsanto Company
Resinox	phenolic	Monsanto Company
Royalene	synthetic rubber*	U.S. Rubber Company
Rulon	fluorocarbons	Dixon Corporation of America
Saflex	polyvinyl butyral	Monsanto Company
Santofome	expanded polystyrene	Monsanto Company
Sicalit	cellulose acetate	Mazzucchelli Cell. s.p.a.
Silastic	silicone rubber	Dow Corning Corporation
Spenkel	polyurethane	Textron Division, Spencer Kellogg Company
Styrofoam	expanded polystyrene	Dow Chemical Company
Styron	polystyrene	Dow Chemical Company
Surlyn	ionomer	E. I. du Pont de Nemours & Company
Sylgard	silicone resins	Dow Corning Corporation
Sylplast	urea resin	FMC Corporation
Synpol	synthetic rubber*	Texas—U.S. Chemical Company
Synvarite	phenolic	Synvar Corporation

Trademark Name	Product	Manufacturer
Tedlar	vinyl fluoride	E. I. du Pont de Nemours & Company
Teflon	poly TFE or FEP	E. I. du Pont de Nemours & Company
Tenite†	various plastics	Eastman Chemical Products
Tetra-Ria	urea resin	National Polychemicals, Inc.
Texin	polyurethane	Mobay Chemical Company
Tipox	epoxy resin	Thiokol Chemical Corporation
Tyril	styrene-acrylonitrile copolymer	Dow Chemical Company
Ultrathene	ethylene-vinyl acetate copolymer	U.S. Industrial Chemicals Company
Unox	epoxy resin	Union Carbide Corporation
Vibrin	polyester	U.S. Rubber Company
Viton	fluorinated rubber	E. I. du Pont de Nemours & Company
Vulcollan	polyurethane rubber	Mobay Chemical Company
Vygen	PVC	General Tire & Rubber Company
Vyram	vinyl resins	Monsanto Company
Zerlon	acrylic-styrene copolymer	Dow Chemical Company
Zetafin	olefin copolymer	Dow Chemical Company
Zytel	nylon	E. I. du Pont de Nemours & Company

* Trademark name for a series of synthetic rubbers, further identified by additional designations.
† General trademark for products.

References

SOURCES OF ADDITIONAL INFORMATION

Many books covering various phases of plastics have been published. Changes have been rapid, tending to make the older books out-of-date. In addition to the books and magazines listed, the annual "Encyclopedia Issue" of *Modern Plastics* magazine is a good source of new developments, which are presented in both the text and the advertisements. House organs, brochures, and various technical releases from manufacturers are useful for new developments and for detailed technical information.

Movie films on plastics subjects, produced mainly by manufacturers and including varying amounts of technical material and advertising, are available commonly for the cost of return postage only. Lists of these can be secured from Manufacturing Chemists' Association, Inc., 1825 Connecticut Avenue, N.W., Washington, D.C. 20036; Society of Plastics Industry, Inc., 250 Park Avenue, New York, New York 10017; and Association Films, Inc., 347 Madison Avenue, New York, New York 10017. Manufacturers can also be contacted directly.

Books on Plastics*

BATEMAN, LESLIE, *Chemistry and Physics of Rubberlike Substances.* New York, Interscience Publishers, John Wiley & Sons, Inc., 1963.

BATTISTA, O. A., *Fundamentals of High Polymers.* New York, Reinhold Book Division, 1958.

BILLMEYER, F. W., *Textbook of Polymer Science.* New York, Interscience Publishers, John Wiley & Sons, Inc., 1962.

BOENIG, H. V., *Polyolefins.* New York, American Elsevier Publishing Company, 1966.

BRYDSON, J. A., *Plastics Materials.* London, Illife Books; Princeton, N.J., D. Van Nostrand Company, Inc., 1966.

GOLDING, BRAGE, *Polymers and Resins.* Princeton, N.J., D. Van Nostrand Company, Inc., 1959.

MARK, H. F., and others, *High Polymers.* A series of Monographs on the Chemistry, Physics and Technology of High Polymeric Substances. 20 volumes to 1964. New York, Interscience Publishers, John Wiley & Sons, Inc.

——, *Encyclopedia of Polymer Science and Technology,* Vol. I—1964, Vol. II—1965. New York, Interscience Publishers, John Wiley & Sons, Inc., 1964–65.

MILES, D. C., and BRISTON, J. H., *Polymer Technology.* London, Temple Press Books, 1965.

MILLER, M. L., *The Structure of Polymers.* New York, Reinhold Book Division, 1966.

MONCRIEF, R. W., *Man-made Fibres,* 4th ed. New York, John Wiley & Sons, Inc., 1963.

MORTON, MAURICE, *Introduction to Rubber Technology.* New York, Reinhold Book Division, 1959.

OKESKY, S. S., and MOHR, J. G., *Handbook of Reinforced Plastics of the SPI.* New York, Reinhold Book Division, 1963.

Reinhold Plastics Applications Series. A series of 30 books on plastics and plastics processing. New York, Reinhold Book Division, 1957–66.

SIMONDS, H. R., and CHURCH, J. M., *Concise Guide to Plastics,* 2nd ed. New York, Reinhold Book Division, 1963.

SMITH, W. M., ed., *Manufacture of Plastics.* New York, Reinhold Book Division, 1964.

Society of the Plastics Industry, *Plastics Engineering Handbook,* 3rd ed. New York, Reinhold Book Division, 1960.

WINDING, C. C., and HIATT, G. D., *Polymeric Materials.* New York, McGraw-Hill Book Company, 1961.

*Only books published in, or a series into, the last ten years.

Selected Periodicals*

Journal of Applied Polymer Science. Interscience Publishers, John Wiley & Sons, Inc., 605 Third Avenue, New York, New York 10016

Journal of Polymer Science. Interscience Publishers, John Wiley & Sons, Inc., 605 Third Avenue, New York, New York 10016

Modern Plastics. McGraw-Hill Publishing Company, 770 Lexington Avenue, New York, New York 10021

Plastics. Temple Press, Ltd., Bowling Green Lane, London, E.C.1, England

Plastics World. Cahners Publishing Company, 221 Columbus Avenue, Boston, Massachusetts 02116

Polymer Engineering and Science. Society of Plastics Engineers, 65 Prospect Street, Stamford, Connecticut 06902

Rubber Age. Palmerton Publishing Company, 101 W. 31st Street, New York, New York 10001

Rubber World. Bill Brothers Publishing Company, 630 Third Avenue, New York, New York 10017

SPE Journal. Society of Plastics Engineers, 65 Prospect Street, Stamford, Connecticut 06902

*Much material relating to plastics will also be found in such periodicals as *Chemical Engineering, Chemical and Engineering News, Chemical Week, Industrial and Engineering Chemistry.*

Index